Domestic Policy Narratives and International Relations Theory

DOMESTIC POLICY NARRATIVES AND INTERNATIONAL RELATIONS THEORY

Chinese Ecological Agriculture as a Case Study

Michael Dalton McCoy

International Scholars Publications
Lanham • New York • Oxford

Library of Congress Cataloging-in-Publication Data

McCoy, Michael Dalton.
Domestic policy narratives and international relations theory : Chinese
ecological agriculture as a case study / Michael Dalton McCoy.
p.   cm.
Includes bibliographical references and index.
1. Agriculture and state—China. 2. Sustainable agriculture—China. 3.
China—Foreign relations—1976- I. Title: Chinese ecological agriculture as a
case study. II. Title.
HD2098.M39   2000   338.1'85l—dc21   99-057989 CIP

ISBN 0-7618-1598-8 (cloth: alk. ppr.)
ISBN 0-7618-1599-6 (pbk: alk. ppr.)

TABLE OF CONTENTS

FOREWORD

Drawing on an interpretive methodology that pays attention to language and metaphor, Dr. Michael McCoy analyzes how sustainable agricultural policy practices articulate political values in the People's Republic of China. He conducted extensive interviews with various participants in the policy process to explain how cultural political values are expressed by and evolve out of Chinese domestic policies. By analyzing the ritual and mythic qualities of these practices, he moves beyond an analysis of their content to a richer regard of how these policies express political values that articulate conflicting domestic concerns. Showing how these values shape not only domestic politics but play a key role in the international politics between China and the United States, he situates his own cultural perspective within the context of the research project. This makes his work especially important for understanding interactions between China and the United States.

McCoy uses key works in international relations to show how this interpretive method can be useful in understanding complexities within the field of international relations. His focus on the language of agricultural policies forms the base of his analysis. Blending this analysis with data gleaned from a variety of sources from China, the United Nations, and the United States, he illuminates how the cultural meanings of these policies shape Chinese political models. His focus on the Chinese practice of employing models to shape local policies shows how Chinese policies evolve in ways that differ from traditional American behavior analyses. Understanding the political implications of these cultural meanings enables McCoy to explain key connections between China's domestic cultural practices and international policies.

Because McCoy's interpretive approach emphasizes cultural similarities and differences, he is able to offer insights on difficulties that arise in Chinese U.S. international situations. Scholars interested in the link between domestic policies and international practices will find this work especially useful in terms of the method of analysis and the cultural insights presented. Those who have substantive interests in agricultural policies will find the work helpful because it reflects on how agricultural policies articulate political values and cultural practices, and those interested in cultural differences between China and the United States will find this work valuable because it offers concrete examples of how those differences can work at the international level of socio-political life.

The interdisciplinary scope of this project gives it broad appeal to those interested in cultural analysis. McCoy's commitment to understanding politics makes the work exciting, because it enables the reader to understand Chinese political culture in a way that respects the intersection between the concerns of political scientists and the concerns of policy makers and participants. What is truly distinctive about this analysis is McCoy's ability to use an interpretive approach to offer reflections that include an analysis of American culture as well as Chinese policy practices. He is part of the new political scientists who critically examine their own cultural values as they examine those of others. The result is refreshingly honest and intellectually engaging.

Eloise A. Buker
Professor of Political Science
Denison University

FOREWORD

This is a book about China and a book about how we know other cultures. The topics of the book are Chinese ecological and sustainable agriculture in their cultural and political context, but in exploring these topics Michael McCoy takes us on a journey through explorations of three interrelated subjects. He examines the development of personal knowledge of the 'other' - his ethnographic journey. He links this to a broader understanding of values and systems of knowledge in the U.S., and he addresses the nature of international understanding and international interaction. The interpretation of symbolic systems in culture links all the diverse themes developed in this text.

China is a prominent topic in today's academic landscape because the massive significance of China can no longer be dismissed by Western scholars. Without repeating the litany of features which are commonly recited in a discussion of China's probable world position in the 21st century I would like to go directly to a focus on the significance of this text for potential readers. My delight in this volume is that I gain insights about myself as a Western observer working to understand China at the same time I explore and learn about the politics of China's ecological agriculture. In this forward I will briefly prefigure some of the themes and ideas you will find in the volume. Above all else I want to recommend that this is a work which is appropriate for general readers, for readers who are interested in the mechanisms of Chinese construction of reality, and for readers interested in how Westerners construct their images of China. McCoy cites a multitude of sources that will make students of political science, history, sociology, anthropology, international relations and Asian studies all feel at home, but his work is also presented in such a clear form that general readers are not left behind.

Chapter 1 introduces the reader to Chinese ecological agriculture and to McCoy's methods and interpretive framework. An especially rich point comes from his self-identification in multi-dimensional intellectual space. In this work McCoy is an American studying China, and a political scientist studying agriculture, but he takes us through a rich body of knowledge which documents the art and science of knowing the 'other'. The reader can expect to complete this work with a major body of fresh knowledge about China in hand. But through McCoy's analysis of the tensions in interpretive space the reader can also expect to complete the work with very important self-insights in hand, and with a provocative assessment of the role of U.S. culture in establishing the systems of knowledge we use to know others. Readers will find this chapter to contain a blend of economic, ethnographic and historical approaches.

In the second chapter McCoy introduces his approach to symbolic interpretation. He explores metaphors, rituals, myths, and his ethnographic experience. The issue which stands out for me in reading this section is the importance of being open to the ideas of the 'other' in any ethnographic context. McCoy describes how he worked to see past his preconceived knowledge, and I find his work very helpful in engaging this struggle.

McCoy examines the strengths of science and the challenges to modern thought of the post-modern critique, and he strives to establish an approach using both foundations. I believe that he ends up clearly in the empirical camp in terms of his actual work, but his review of these paradigms leads the reader to the self-evaluation which is perhaps the strongest contribution of this work. In the best tradition of modern ethnography, McCoy tells us his story of learning. He provides actual conversations, details the interpretive mistakes he made, and discusses the subsequent information he gained which allowed him to see why he was missing the Chinese cognitive point. By extension the reader can begin to explore her/his own assumptions and can identify with and practice techniques of knowing the 'other' which include a sensitivity to the structures of knowledge of that 'other'. I would define McCoy as an ethnological political scientist.

The third chapter provides a history of ecological agriculture followed by a review of the Chinese philosophy of ecological agriculture. This gives the reader a foundation for understanding Chinese ideas and discourse today. In the fourth chapter McCoy interprets Chinese metaphors related to understanding ecological agriculture. He introduces the reader to the use of model farms in China and to a Chinese model farmer. This sets the stage for the fifth chapter introduction of model-following as an important aspect of Chinese psychological space. Model-following, which is not a strong aspect of Western culture, is a central part of Chinese enculturation and is a metaphor applied in many aspects of Chinese thought.

Among the important topics McCoy reviews are issues of whether or not farmers have input in the development of Chinese government policy, how change is implemented in China, and how China differs from the U.S. in conceptualizing 'the future' and 'the past'. Readers will enjoy meeting the 'model farmer' and finding out about his work. The use of 'models' - individuals, groups, practices - is central to the Chinese practice of development and is very foreign to the American practice. In the U.S. our cultural icons develop their own positions, and are usually seen as having no relationship to government. In China the government uses the establishment of model projects as one of the regular approaches to the implementation of new policy. In developing points like this, McCoy makes this book a very rich text for all students of Chinese culture.

This volume shares a range of examples and interpretations which help the reader grasp the complexity of decision making in China and the symbols and themes which provide common foundations for Chinese understanding. His goal in this work is to take the quality of 'enigmatic' out of our understanding of Chinese decision-making, and his presentation works well. I believe that Chinese readers will also come away from this text with a fresh sense of Western intellectual patterns and of some of the foundations of failed communication between members of the two cultural traditions. In concluding Chapter 5 McCoy develops a very important exploration of some of China's major identity constructs in presenting itself to the world today, and opens discussion regarding the possible images of China in the next millennium. Among the issues he discusses the reader will find a challenging critique of China's investment in maintaining an image as a third-world country.

The final section of the volume focuses on what one needs to know about a culture's internal framework (domestic policy) in order to understand its international practice. While the focus is on understanding China, the text is just as rich in looking at the structure of our own culture. The text raises the issue of the variety of perspectives needed to work across cultural boundaries. As Westerners we must understand that our existing knowledge structure gets in the way of understanding other cultures, so we study China, trying to decode the systems of meaning of the Chinese and to apply those systems, while seeing ourselves as guided to different perspectives by the structures of our own knowledge. Each cultural system has a body of knowledge based on history, culture and experience. Cross-cultural communication is not likely to be effective unless people recognize the differences between cultures, and this is a difficult task. In applying this scheme McCoy looks at China's U.N. behavior and provides us with a fresh construct for understanding China's actions. I can't say whether or not he is correct, but this revision is very challenging! This is the political science and ethnoscience chapter, and provides a complex, thought-provoking conclusion.

This volume contributes to the success of cross-cultural communication by providing clear examples of the complexity of international relations. The reader can add these materials to his/her own body of experience in order to develop the skills needed to hear past the structure of Western knowledge. One of the most striking interpretations McCoy offers is that there may be a sense in which Chinese democracy can be recognized as quite strong. He develops this idea through an exploration of the degree to which the Chinese government listens to the will of the people - in other words there is the structure of democracy as a voting process and the structure of democracy as an enactment of the will of the people. As I integrate this idea with my current knowledge of the buying of the American voter I have a frightening reevaluation of the nature of democratic futures.

In this forward I have touched on a few of the issues which make this work exciting. I believe that the richness of this work will keep it in readers' hands for many years to come, and I encourage readers with many different interests to press on and encounter the delight of this volume.

Pat Wenger
Professor of Anthropology
Humboldt State University
Arcata, California

ACKNOWLEDGMENTS

My deepest appreciation and thanks goes to my family and especially to my wife Sherry, who gave as much towards this work as I did. I would also like to thank James Mayfield, Eloise Buker, Dvora Yanow, Mary Hampton, Yanqi Tong, FuNing Zhong of Nanjing Agricultural University, Neal Bergstrom, and the Mariner S. Eccles Foundation for their for their encouraging and unconditional support and assistance. I would also like to thank the many people, some of whom I met only on phone calls, who gave so generously of their time and knowledge. Thank you.

CHAPTER 1

INTRODUCTION

Chinese agriculture is facing some significant problems in terms of diminishing resource availability, declining resource quality, and increasing production demands as a function of rapid population growth. Structural changes such as the household responsibility system, increasing marketization, price reforms, land reclamation, and high technology inputs (high yield seeds, fertilizer, pesticides, mechanization) have done much to address these problems. In light of this, grain production has not continued to grow consistently at expected levels, the majority of farmers are still only producing enough to meet household needs, adoption of market mechanisms and high technology inputs is slow, and resource depletion is becoming an increasingly significant problem.

Chinese ecological agriculture (CEA) is a reaction to the excessive and damaging effects of China's experiment with the 'Green Revolution' of the last thirty years.[1] CEA is a government sponsored project, that since 1986 has been encouraging farmers to experiment with increased integration of animal, plant, and soil resources which leads to an increased use of natural fertilizers with the use of chemical fertilizers and pesticides in a minor roll as a supplement. It is a uniquely Chinese approach to their agricultural dilemma that is a product of domestic and international influences. CEA addresses many of China's

---

1. The term 'Green Revolution' refers to the use of high yielding varieties of seed and the intensified use of irrigation and chemical fertilizers necessary for the new seeds to work. This technology came into being in the 1960s and is the core of conventional, modern agricultural systems.

environmental and agricultural production problems. While working within the existing constraints of small family plots and limited mechanization, CEA has resulted in increased yields in food and cash crops, improved soil quality, and in the right setting relatively easy adaptation by farmers.[2]

The efforts of CEA are interesting because they are a direct reaction to the wholesale application of Western ideas and technology to the process of Chinese development. How the Chinese discuss and implement this approach to agricultural development reflects more than just a creative technical approach to current agricultural issues; it also reflects domestic cultural cues which are an integral part of China's political processes and identity.

Understanding these cues through interpretive analysis, in the context of policy implementation, can give us a policy-based understanding of a Chinese government approach to goal setting, problem definition/resolution, and cooperation. These interrelated facets of development are a part of China's contemporary cultural identity, the understanding of which has ramifications for observers of China's domestic and international activity.

Purposes and Objectives

The primary purpose of this study is to examine CEA as an example of domestic politics in order to address the levels-of-analysis issue in the international cooperation literature. Specifically, I seek to accomplish the following goals: (1) to show the usefulness of implementation analysis for learning about domestic approaches to cooperation and about the political incorporation of new ideas, (2) to articulate an interpretive approach to the analysis of CEA and its implications for the analysis of international cooperation, and (3) to apply the findings of my CEA analysis to the current debate on

---

2. Cheng Xu, Han Chunru, and Donald Taylor, "Sustainable Agricultural Development in China," *World Development* 20, no. 8 (August 1992); T.C. Tso, *Agricultural Reform and Development in China* (Beltsville, MD: Ideals, Inc. 1990).

international cooperation. The perspective offered by policy implementation studies involves a discussion of the process of politics and the meanings which actors give to a policy and its outcomes. In this context meaning refers to how the participants themselves interpret various concepts and actions. Meaning is important because it shapes and is shaped by future actions. Implementation is about an understanding of who the key actors are, their values, their perceptions about self, other, and the policy. It is also about the goals and intentions of the policy, the available resources, the control of those resources, and the cooperation necessary to achieve the desired goals. What cooperation means domestically can add a different dimension to the analysis of international behavior.

From an interpretive perspective "the policy process may be seen as a 'text' through which members of a polity tell themselves who they are and what they value."[3] I will use an interpretive approach which conceptualizes the processes and artifacts of implementation as symbolic language, objects, and acts.[4] People who were interviewed for this study were Chinese farmers, village leaders, county officials, provincial officials, university researchers, American researchers, and international agency officials. CEA will be analyzed for meanings that are given to and taken from it and for the emergence of cultural themes.

Meanings that have importance domestically are also carried over into the international sphere and therefore have import for the analysis of international relations. The emergent cultural themes are relevant for adding to our understanding of the interpretations which China gives to its cooperative relationships and likewise the meanings which Americans give to the Chinese development experience. The disparity between these two perspectives (i.e., meanings) is the basis for reexamining current academic discussions about international cooperation. The meanings given to the concept of cooperation and

---

3. Dvora Yanow, "The Communication of Policy Meanings: Implementation as Interpretation and Text," *Policy Sciences* (Spring 1994): 43.

4. Ibid.

4

given to the experiences of cooperation cannot be assumed to be shared and, as such, neither can the goals or objectives of cooperation.

Theoretical Issues

My purpose in this section is to explain what the current concepts and dilemmas are in the analysis of state behavior domestically and internationally and how I approach these problems from a different perspective. I will begin with a discussion of the issues from an international relations perspective and move to a domestic perspective using literature that applies the relevant conceptual frameworks to issues in China. This will serve two purposes: (1) to show how the analysis of international relations is largely a conceptual issue and (2) to keep the discussion focused on Chinese politics and processes.

There has been a long-standing debate in international relations regarding which level of analysis, international or domestic, is more fruitful for understanding state behavior. At this point in time it seems to be a draw, those from each perspective acknowledging the importance of the other, but neither can theoretically reconcile the other.[5] Harold Jacobson's and Michel Oksenberg's book, *China's Relations with the IMF, the World Bank and GATT*, is a useful case study for examining the levels-of-analysis debate and the concept of cooperation.[6]

---

5. For a review and sources in the levels-of-analysis debate in international relations studies recent articles by Alexander Wendt and Walter Carlesnas are especially good. They analyze standing issues in the debate and advance it through the precepts of structuration theory. Their work in moving this debate beyond a conceptual deadlock has been met with interest, but skepticism remains due to the lack of a clear methodological agenda. Alexander Wendt, "Levels of Analysis vs. Agents and Structures," *Review of International Studies* 18 (1992): 181-5; Wendt, "The Agent-Structure Problem in International Relations Theory," *International Organization* 41, no. 3 (Summer 1987): 335-70; Wendt, "Anarchy is What States Make of it: The Social Construction of Power Politics," *International Organization* 46, no. 2 (1992): 391-425; Walter Carlsnaes, "The Agent-Structure Problem in Foreign Policy Analysis," *International Studies Quarterly* 36, no. 3 (1992): 245-270.

6. Harold K. Jacobson and Michel Oksenberg, *China's Participation in the IMF, the World Bank, and GATT* (Ann Arbor: University of Michigan Press, 1990). The current literature on international cooperation has largely been devoted to the concept of international regimes, their function, formation and decline in international relations.

In this book international cooperation is addressed as a function of China's process of admission into these Bretton Woods organizations. Jacobson and Oksenberg use Robert Keohane's theory of international regimes, whereby the structure of rules created by the Bretton Woods institutions is the framework within which China's cooperation can be gauged. Keohane's theory posits that international structures shape state behavior and that the outcome of cooperation is "mutually beneficial policy coordination."[7]

Keohane defines cooperation as an effort by states to coordinate economic policies; it is "mutual adjustment."[8] Cooperation is a distinct middle ground between the concepts of harmony and discord. Harmony is the automatic attainment of goals, and is distinct from discord which is the belief that the policies of other states are hindering the attainment of one's own policies.[9] Cooperation as such is an essential part of international behavior, and is the basis for the development of and adherence to international regimes. It is a state

---

Ernst S. Haas, "Regime Decay: Conflict Management and International Organizations," *International Organization*, 37, no. 2 (Spring 1983): 189; Peter Haas, "Do Regimes Matter? Epistemic communities and Mediterranean Pollution Control," *International Organization*, 43, no. 3 (Summer 1989): 377; Stephan Haggard and Beth A. Simmons, "Theories of International Regimes," *International Organization*, 41, no. 3 (Summer 1987): 491; Stephen D. Krasner,ed., "International Regimes," *International Organization*, 36, no. 2 (Spring 1982); Friedrich Kratochwil and John G. Ruggie, "International Organization: A State of the Art on an Art of the State." *International Organization*. 40, no. 4 (Fall 1936): 753; Vinod Aggarwal, "The Unraveling of the Multi-Fiber Arrangement 1981: An Examination of International Regime Change," *International Organization*, 37, no. 4 (Spring 1983): 617; Benjamin Cohen, "The Political Economy of Trade,"*International Organization*, 44, no. 2 (Spring 1990): 261; Peter Cowhey, "The International Telecom-munications Regime: the Political Roots of Regimes for High Technology," *International Organizations* 44, no. 2 (Spring 1990): 169; Jack Donnelly, "International Human Rights: A Regime Analysis," *International Organization*, 40, no. 3, (Summer 1986): 599; Oran R. Young, "International Regimes: Problems of Concept Formation," *World Politics* (Summer 1980): 331; Oran R. Young, "The Politics of International Regime Formation: Managing Natural Resources and the Environment," *International Organization*, 43, no. 3 (Summer 1989): 377.

7. Robert O. Keohane, *International Institutions and State Power* (Boulder: Westview Press, 1989), and Keohane, *After Hegemony: Cooperation in the World Political Economy* (Princeton: Princeton University Press, 1984).

8. Keohane, *After Hegemony: Cooperation and Discord in the World Political Economy*, 12.

9. Ibid., 51.

6

behavior that needs to be understood and facilitated for maintaining international political stability.

From this perspective, understanding China's international behavior and its desire to cooperate is a function of how closely China follows the existing rules and norms of the international monetary system.[10] Jacobson and Oksenberg's questions are necessarily limited to whether or not China has accepted and follows the established Bretton Woods rules, or if the Chinese have pushed for changes in those rules that would be more favorable to their own "traditions and practices."[11] They do not ask whether or not such rules were efficacious in solving China's domestic problems. In answering these questions Jacobson and Oksenberg accept Robert Keohane's assumptions about international institutions and regimes.

Keohane's approach and assumptions are based on the important role of institutions and regimes in international relations. International institutions and regimes are the result of a need for cooperation among states in an anarchical world. "Interdependence is high in the contemporary era due to the progress of science and technology; avoiding harmful and destructive actions and achieving mutually beneficial goals require that states adjust and coordinate their policies."[12] Institutions such as the World Bank "create and maintain networks that gather information" which are important for consensus, help overcome "political market failure" and can "lead to a convergence of values."[13] The primary assumption embodied in this approach is that states are rational actors that need to cooperate in order to achieve their own economic goals.

---

10. Samuel Kim makes a similar assessment regarding China's participation in the United Nations in Samuel Kim, *China the United Nations, and World Order* (Princeton: Princeton University Press, 1979).

11. Jacobson and Oksenberg, 5.

12. Ibid., 7.

13. Keohane, *After Hegemony*, 90.

State behavior, according to Keohane, must be understood within the context of international systemic constraints (i.e., the world economic system). The goal for Keohane is to understand how systemic constraints affect state behavior, with the belief that cooperation will not "necessarily foster humane values, but effective coordination of policy would often help."[14] Domestic factors are important but are considered as constants in his systemic analysis. As dictated by microeconomic theory domestic politics are reconceptualized into the individual attributes of rationality and egoism. A state's rationality for action is based on consistently ordered preferences and the calculation of costs and benefits. Egoism refers to a state's utility functions, its ability to define self-interests by consistently choosing between preferences that are ordered by rationality.

Rationality and egoism, ordering preferences and choosing preferences, are subjective and independent acts that vary from state to state. For Keohane, the result of this confusing and individual focus on behavior is to treat the rationality and self-interests of the state as analytical constants; they are assumed. The focus must shift to a macroview of international relations in which variations in [state] behavior are accounted for through variations in the international economic system. "Without the assumptions of egoism and rationality, variations in (state) behavior might have to be accounted for by differences in values. In that case, analysis would revert to the unit level, and the parsimony of systemic theory would be lost."[15]

Using Keohane's theory of regimes to guide their analysis of Jacobson and Oksenberg maintain a macrolevel of analysis that seeks to explain Chinese

---

14. Keohane, *International Institutions and State Power*, 11.

15. Ibid., 25-27. Keohane does observe that "accepting rational-egoist assumptions involves taking seriously a purely hypothetical notion of rationality that does not accurately model actual processes of human choice (p.29)." Yet it does have three values: 1) "it simplifies our premises, making deductions clearer," 2) maintains a focus on the international system, 3) it allows Keohane to critique mainstream international relations theory from within its own assumptions (p.29).

8

international behavior by examining international structures. In their case study they examine the rules of the Bretton Woods regime as an international structure for cooperation, and so consider that China's acceptance of and following of these established rules and norms for cooperative behavior are measures of Chinese international cooperation. Given the West's concerns about China as a Communist regime and its open support for the New International Economic Order (NIEO), these questions and the approach used are not altogether unwarranted. In showing that China has accepted the existing rules of the Bretton Woods regime and therefore seems to be cooperative, the focus of their analysis shifts to the future. The institutions of the Bretton Woods regime can "play an influential role in easing the process (of cooperation) forward and institutionalizing it."[16]

Oksenberg and Jacobson believe that the possibilities for a strengthening or weakening of this cooperative relationship are rooted in Chinese domestic politics. They view the uncertain outcome of China's domestic political and economic reform issues as potential detractors to a convergence of strategic interests between the World Bank (International Bank for Rural Development or IBRD) and China. For Jacobson and Oksenberg domestic factors, such as loss of control by the central government, or the Chinese concern about the corruptive influences of Western ideas could potentially weaken China's reputation as a reliable IBRD member. These factors might push China to voluntarily minimize interaction with Western financial and trading institutions.[17] This of course would be a setback to efforts to maintain a politically stable China through increased economic interdependence.[18]

---

16. Jacobson and Oksenberg, 165.

17. Ibid., 160-164.

18. "China could prove quite disruptive if its expansion occurred without concern for established trading patterns, or the process (of economic growth) could be derailed if fears in the developed world brought on rampant protectionism. Thus, the ability to absorb China into the world economy will be substantially shaped by the perceptions and expectations of the Chinese and their

These conclusions assume that Chinese and World Bank interests are relatively static and do not allow for state cr institutional learning which might modify the interests and the meanings which each actor gives to the relationship. A more appropriate way to understand the Chinese domestic political arena and its impact on international behavior is to see domestic issues not as a set of assumptions to be bracketed for the sake of theoretical tidiness or as inexplicable detractors to cooperation but as an integral part of the larger reform process. Implementing economic reforms and experimenting with capitalist methods of resource allocation, while maintaining socialist political control, are inherently dialectical political processes that demand at least the equal consideration of domestic politics as well as international pressures and opportunities.

The analysis by Jacobson and Oksenberg highlights the conceptual dichotomy between international cooperation and domestic politics; that by focusing on the international structure as the context for understanding state behavior, domestic issues become secondary. Yet, as they correctly point out, it is at the domestic level of action in which the short-term effects and long-term consequences for international cooperation take shape. It is the limited ability of Keohane's theory to account for domestic politics that also limits Jacobson and Oksenberg's analysis. Keohane himself acknowledges the shortcomings of his theory for explaining the normative, domestic aspects of state behavior, specifically changes in beliefs which lead to changes in a state's interests.[19]

Stephan Haggard and Beth Simmons have categorized the theories which consider the more normative, domestic aspects of state behavior as cognitive theories and place them in contrast to the structural-functional theories such as

---

trading partners concerning global economic trends and the consequences of Chinese behavior in international affairs." Ibid., 13.

19. Keohane, *International Institutions and State Power*, 171.

10

Keohane's neo-institutionalism.[20] Cognitivists are critical of functional theories as being too static. One weakness in functional theories is the assumption that international structures which shape state behavior are not very dynamic over time and that there is a high degree of continuity in state's interests. Functional theories

> are incomplete, since they ignore changes taking place in consciousness. They do not enable us to understand how interests change as a result of changes in belief systems. They obscure rather than illuminate the sources of a state's policy preferences.[21] Cognitivists are concerned with explaining cooperation through ideology, the values of actors, the beliefs they hold about the interdependence of issues, and the knowledge available to them about how they can realize specific goals. Cooperation is affected by perception and misperception, the capacity to process information, and learning.[22]

---

20. Realist theories are also referred to by Keohane as rationalistic theories. Theories that are concerned with the normative aspect of state behavior are less easily labeled. As mentioned, Haggard and Simmons use the term "cognitive"; Keohane prefers the term "reflective." For this portion of the paper, I will continue with the label of "cognitive" as distinct from my later use of the term "interpretive" because their use of the term "cognitive" does not pursue the question of meaning as a hermeneutic problem, but relies on the interpretation provided by the construction of a model or inferred from the analysis of behavior. The issue is important because it reflects two different philosophical conceptions of what is science and will be taken up later in this chapter. See Stephan Haggard and Beth Simmons, "Theories of International Regimes," *International Organization* 41, no.3 (Summer 1987): 491-517; Keohane, *International Institutions and State Power*, 161.
See also Friedreich Kratochwil and John G. Ruggie, "A State of the Art on an Art of the State," *International Organization*, 40 (1986): 753-776; Emanuel Adler, ed., *Progress in Postwar International Relations* (New York: Columbia University Press, 1991); Ernst B. Haas, "Words Can Hurt You; or, Who Said What to Whom about Regimes," *International Organization*, 36, no. 2 (Spring 1982): 207-244.

21. Keohane, *International Institutions*, 171. See also Robert W. Cox, "Social Forces, States and World Orders: Beyond International Relations Theory," 204-254, and Richard K. Ashley, "The Poverty of Neorealism," 255-300, in Robert O. Keohane, ed., *Neorealism and Its Critics* (New York: Columbia University Press, 1986).

22. Haggard and Simmons, 510.

The effect is to see cooperation as a dynamic process over time and not to assume that different actors "will respond similarly to the same structural constraints and opportunities; much depends on past history, knowledge, and purpose."[23]

Cognitivists, however, do not have all of the answers. The primary criticism of the cognitivist school is its 'lack of a clear reflective research program that could be employed by students of world politics."[24] They cannot make generalizable statements as structuralists can, nor can they "describe clearly how power and ideas interact."[25] In spite of the differences between these two schools of thought, proponents of both recognize the valuable contribution which the other makes to international relations analysis and they advocate "a synthesis," or a bridging of the analytical gap between structure and process.[26]

Using the cognitive approach it cannot be assumed that actors share the same values, beliefs, or perceptions of international cooperation with one another. However, similar to systemic analysis, one of the shortcomings of the cognitive approach is how to understand the specific reciprocal effect of domestic politics and international interaction. The cognitive approach does show the need to acknowledge that each political actor operates within different historical and cultural boundaries, but it does not offer any strong indication of how to look at the domestic aspect of international cooperation.

Through the processes of policy implementation within China, the domestic issues become symptomatic of China's incorporation of Western ideas and technology. These are the effects of cooperation. It is during the process of policy implementation that Western ideas and intentions meet with Chinese ideas

---

23. Ibid., 511.

24. Keohane, *International Institutions and State Power*, 173.

25. Haggard and Simmons, 512.

26. Keohane, *International Institutions and State Power*, 174. Robert D. Putnam makes a specific attempt to do this in "Diplomacy and Domestic Politics: the Logic of Two-Level Games," *International Organization* 42, no. 3 (Summer 1988): 427-460.

12

and intentions. The outcomes of implementation help shape the path of development and give meaning to concepts such as learning, perceptions, ideology, and interests. Understanding what meaning these concepts have for the Chinese should be an integral part of studying domestic development and international cooperation.

The perspective offered by policy implementation studies involves a discussion of the process of politics and the meanings which actors give to a policy and its outcomes.[27] *Bureaucracy, Politics, and Decision Making in Post-Mao China*, by Kenneth Lieberthal and David Lampton, illustrates the intricacies of policy implementation in China and highlights important processes and relationships.[28] One of the primary conclusions is that efforts to reform China's economic structure have met with difficulties because the existing organizational values have perpetuated the same bargaining behaviors which the

---

27.   The book by Jeffrey L. Pressman and Aaron Wildavsky, *Implementation* (Berkeley: University of California Press, 1984), is the seminal work within implementation studies. For an excellent overview of where the field stands today and the directions that it is headed in, see Dennis J. Palumbo and Donald Calista, *Implementation and the Policy Process: Opening Up the Black Box* (New York: Greenwood Press, 1990).

The following provide examples of implementation studies applied to the area of economic development: Derick W. Brinkerhoff and Arthur A. Goldsmith, "Promoting the Sustainability of Development Institutions: A Framework for Strategy," *World Development Journal* 20, no. 3 (1992): 369-384; Marc Lindenberg,"Making Economic Adjustment Work: The Politics of Policy Implementation," *Policy Sciences* 22 (1989): 359-394; John W. Thomas and Merilee Grindle,"After the Decision: Implementing Policy Reforms in Developing Countries," *World Development* 18, no. 8 (1990): 1163-1181; John W. Thomas and Merilee Grindle,"Policy Makers, Policy Choices, and Policy Outcomes: The Political Economy of Reform in Developing Countries," *Policy Sciences* 22 (1989): 213-248.

28. Kenneth Lieberthal and David Lampton, *Bureaucracy, Politics,and Decision Making in Post-Mao China* (Berkeley: University of California Press, 1992). For other implementation studies that deal specifically with China see also David Lampton,ed., *Policy Implementation in Post-Mao China* (Berkeley: University of California Press, 1987); Xu Cheng, Han Chunru, and Donald Taylor, "Sustainable Agricultural Development in China," *World Development* 20, no. 8 (August 1992): 1127-1144; Zhu Ling, "The Transformation of the Operating Mechanisms in Chinese Agriculture," *The Journal of Development Studies* 26, no. 2 (January 1990): 231; Kristen Parris, "Taxing the Private Sector in China: Implementing Policy Reform," Paper presented at the Annual Meeting of the Western Political Science Association, Pasadena, California, March 1993; W.R.Lavely, "The Rural Chinese Fertility Transition: A Report from Shifang Xian, Sichuan," *Population Studies* 38, no., 3 (1984): 365-384; Jonathan Unger, "Bending the School Ladder: the Failure of Chinese Educational Reform in the 1960's," *Comparative Education Review* (June 1980): 221-237.

reforms of decentralization and marketization were intended to change. The point that the nature of relationships has remained the same in spite of decentralization is emphasized in the following quote from the domestic Chinese news.

The current outstanding problem is that enterprise management mechanisms haven't been changed ...the basic economic relations between the government and enterprises, between various departments, and between the central and local authorities have yet to be completely straightened out.[29]

Recent bureaucratic reform efforts in the People's Republic of China have been implemented with the intention of improving economic efficiency and performance by placing decision-making responsibilities at lower administrative levels. Structurally the reforms have given extensive administrative powers to the provinces which were previously held by the central government. Likewise central cities (key economic centers) and the counties have been given increased administrative autonomy vis-á-vis the provinces. Although the increase of economic performance has occurred to a considerable degree, the creation of multiple centers of power has led to increased inefficiencies in decision making and policy implementation.

One of the primary reasons for decentralizing administrative responsibilities was to break the traditional reliance on vertical and horizontal lines of bureaucratic responsibility. (Vertical lines are connections to central decision makers, and horizontal lines link organizations with overlapping powers of authority.) The outcome of decentralization was intended to increase the ability of the domestic actors, such as provincial, city, and county leaders, to make autonomous decisions. Unfortunately, these lines of responsibility are not only structural components of the Chinese bureaucratic system which can be altered with a single policy, but they represent deeply imbedded values, primarily those of organizational autonomy, and integrated administrative action or centralized

---

29. Xinhua domestic service, in FBIS January 15, 1992, 40.

14

political control.[30] Both sets of values, organizational autonomy and centralized control, must be considered as integral components of any structural changes and their resulting outcome. These values are fundamentally different from Western management values of organizational interdependence, and decentralized political control, and therefore give rise to meanings of action, such as the notion of administrative efficiency, which are also different from those in the West.

Kenneth Lieberthal, using a model he calls `Fragmented Authoritarianism' "argues that authority below the very peak of the Chinese political system is fragmented and disjointed."[31] This fragmentation has resulted in increased bargaining among domestic actors, and it is the concern of the authors to understand the "conditions under which bargaining does and does not occur."[32] Lieberthal and his colleagues firmly conclude that the bureaucratic bargaining as it now exists is an inefficiency and that bargaining as a behavior must be institutionalized into more clearly delineated lines of responsibility and norms for conduct.[33]

'Fragmented Authoritarianism' does provide an understanding of how the structures and processes result in bargaining, but the expressed author's omission of values as a necessary aspect of understanding current bureaucratic behavior leaves the question of what the concepts of efficiency and bargaining mean to the Chinese. Indeed there may be a recognized need by the Chinese and others for more clear lines of responsibility; however different notions of efficiency and bargaining will lead to different types of relationships. The values of

---

30. Cao, Dexiang, *Public Administration in Post-Mao China (1978- 1988): A Theoretical Analysis Through Rosenbloom's Model*, Ph.D. dissertation, State University of New York at Albany, 1990.

31. Kenneth Lieberthal and David M. Lampton, *Bureaucracy, Politics, and Decision Making in Post-Mao China* (Berkeley: University of California Press, 1992), 8.

32. Ibid., 9.

33. These findings mirror the findings in an earlier book, Kenneth Lieberthal and Michel Oksenberg, *Policy Making in China: Leaders, Structures, and Processes* (Princeton: Princeton University Press, 1988).

organizational autonomy and integrated administrative action become the context within which the bureaucratic actors at all levels of government give meaning to the decentralization of authority and resources. Decentralization has created more centers of power, each wanting to be autonomous in terms of resources and at the same time trying to figure out how to maintain an integrated administrative structure.

The current reforms as described by Lieberthal seem to be institutionalizing this `inefficient' bargaining behavior precisely because the values of organizational autonomy and integrated administrative action are mutually reinforced through decentralization by creating multiple centers of power. Bargaining has occurred among new authority centers in an effort to compete for resources, to remain organizationally autonomous, and to establish new political relationships for integrated administrative action.

With these values as underlying bargaining behavior, the focus of the problem shifts from how to institutionalize bargaining behavior so it becomes more 'efficient' (yet for the Chinese this behavior seems to mean efficiency)[34] to what do these underlying values mean to the Chinese? Is bargaining the acceptable way of conducting business or is it a temporary necessity brought on by the confusion of the reforms? What would be considered by the Chinese to be an efficient administrative organization? Answering these types of questions through the incorporation of the interpretation of cultural values into the analysis of political processes and relationships involves a discussion of meanings.[35]

---

34. Cao, 84.

35. In her discussion about the implementation of policy in the United States, Barbara Freeman discusses fragmented authority and bargaining as characteristics of the American system. Within our democratic system Americans often consider this to be a virtue in a process that allows multiple voices and values to become a part of policy formation. However, the conclusion by Lieberthal is that in China it is inefficient government because it deviates for the original goal of "substituting market forces for bureaucratic management" (p.12). This analysis overlooks the fact that market forces are made up of the bargaining among different groups of people competing for resources and commodities. From an American perspective of how business ought to be conducted the Chinese system may seem to be very inefficient, but what would the Chinese consider to be more efficient and why? See Barbara Freeman, "Implementation and Madisonian

Meaning is about culture. The analysis of culture is "not an experimental science in search of law but an interpretive one in search of meaning."[36] Culture is a way of being, of defining oneself, and of interpreting the world. Culture, whether it is embedded in a specific organization or in social norms, influences all human behavior and underlies any activity which is undertaken. It is not enough to understand the nature of political relations in a specific issue area or to know how the societal nature of those relationships impacts individual behavior. The meaning which individuals ascribe to those relationships will guide their future actions. "Culture is not a power, something to which social events, behaviors, institutions, or processes can be causally attributed; it is a context, something within which they can be intelligibly ... described."[37]

For many cultural theorists in political science the problem is always how to "operationalize" the concept of culture, and their inability to construct a theory that causally accounts for change or learning.[38] Michael Thompson, Richard Ellis, and Aaron Wildavsky in *Cultural Theory* have developed a group/grid model that they believe addresses these issues.[39] In their group/grid analysis five

---

Government," in Dennis J. Palumbo and Donald J. Calista,eds., *Implementation and the Policy Process: Opening Up the Black Box* (New York:Greenwood Press, 1990): 39-50.

36. Clifford Geertz, "Thick Description: Toward an Interpretive Theory of Culture," *The Interpretations of Cultures* (Basic Books Inc., New York, 1973): 3.
    For general discussions of the concept and analysis of culture, see Ronald Chilcote, "Theories of Political Collectivity and the New Person" in Chilcote, *Theories of Comparative Politics* (Boulder: Westview Press, 1981); Linda Smircich, "Concepts of Culture and Organizational Analysis," *Policy Sciences* 19 (1986); Stephen Ott, *The Organizational Culture Perspective* (Pacific Grove,CA:Brooks/ Cole Publishing, 1989).

37. Ibid., 15.

38. Harry Eckstein in "A Culturalist Theory of Change" notes that culturalist theories create an expectation of continuity and therefore can not account for change or learning. He outlines a culturalist theory of political change based on Talcot Parson's action frame of reference, and concludes that the problems of testing and operationalizing the concepts remain a problem.
Harry Eckstein,"A Culturalist Theory of Change," *American Political Science Review* 82, no. 3 (September 1988), 789-804.

39. Michael Thompson, Richard Ellis, and Aaron Wildavsky, *Cultural Theory* (Boulder: Westview Press, 1990).

ways of life (hierarchical, egalitarian, fatalism, individualism, and autonomy) are described, replete with concomitant value sets and anticipated behavior based on the interaction of these five categories of living.

According to the authors it is because these categories are "formed from dimensions [multiple categories such as biological, geographical, or technical] rather than derived ad hoc from observation" create a logical coherency because all five are "mutually exclusive and jointly exhaustive."[40]   Based on their theoretical typology the prediction of new social preferences and change becomes for them an easy matter.   Using a grid/group model, the authors claim to categorize all people into four groups and define how each type will deal with risk in generating social preferences and how they will change based on their reactions to surprises (unanticipated outcomes of action).   For example, according to the model those people that fall into the category defined as 'individuals' see risk as an opportunity and are surprised by their failures.   Their response in a risk situation will be to act in a pro-active manner; if the results of that action are a failure, then this group will be surprised and reassess the situation.   Their reassessment will result in a form of change in group preferences.

For the authors the Cultural Revolution in China is easily explained with this theory.

> Viewed through a cultural lens, there is nothing particularly novel about the Cultural Revolution, except its extremity, a fact that is due to egalitarian's [the theory's classification of Mao and his supporters for the Cultural Revolution] ability to seize near total power

and their belief in a "pure inside and outside" that encourages the search for hidden enemies.[41]   This egalitarian behavior is explained by the fact (a fact as established by the particulars of their model) that when people defined by the

---

40. Ibid., 14.

41. Ibid., 229-231.

18

model as egalitarians "construct a way of life combining strong group boundaries [for group identity] with weak prescriptions [for behavior] they must reconcile collective choice with individual autonomy."[42]   In the case of the Cultural Revolution the need to maintain strong group boundaries won out. A simple explanation for a complex social and political event.

In contrast to *Cultural Theory* where the focus of analysis is on the interaction of different group types, with given values as predictors of expected behavior, other cultural theorists are concerned with the process of value formation as an analytical variable for understanding political behavior. Lowell Dittmer and Samuel Kim's recent book, *China's Quest for National Identity*, relies on psychological theories of identity development to fulfill the objective of understanding organizational development.[43]

Dittmer and Kim's book is interesting because it asks the question, "How is China's identity changing and emerging?" Rather than using a static typology, *China's Quest for National Identity* is based on the idea of symbolic interactionism within Erik Erikson's model of psychological development to describe how China sees itself and its place in the world.

> The substantive content of national identity is the state and that the state defines itself by what roles it plays - [*sic*] by self-categorization in alignment with positive reference groups and in opposition to negative reference groups - by its performance in the international arena.   But the state defines itself not only behaviorally but essentially - by what it 'is' as well as what it 'does.'[44]

A state's essence, what a state "is," involves an "ensemble of symbols collected to represent the principles on which the group was founded and on the

---

42. Ibid., 262.

43. Lowell Dittmer and Samuel S. Kim, *China's Quest for National Identity* (Ithaca: Cornell University Press,1993). See page 3 for their discussion of Eric Erikson.

44. Ibid., 17.

basis of which its members have contracted to live together."[45] Symbols refer to actual events but go beyond the event to "express and convey emotion."[46] Dittmer and Kim use one definition of a myth to explain how political leaders life to state symbols. For these authors, myths are the reinterpretation of historically grounded cultural symbols to justify the present reality through an "invented" explanation; a form of false consciousness.[47]

When symbols and myths are combined together within the psychological stages of development as defined by Erikson, a model of political development is created which defines what types of cultural land marks should be used as standards of development, and how the meanings and growth of those landmarks should ideally mature over time. Within this construction created by Dittmer and Kim, national identity has a "normal" developmental pattern that is desirable. However, when a nation's sense of identity is in crisis an abnormal pattern of development emerges. It is implied by these authors that China is in or close to such an abnormal possibility through their frequent analogies that compare modern China to Nazi Germany.[48] A more normal pattern of development would be more democratic, market-oriented, and cooperative. How these concepts used in this model are defined and by whom is the issue that I believe must be addressed more fully.

---

45. Ibid.

46. Ibid.

47. Ibid., 18.

48. Nazi Germany, used as an analogy for contemporary China and as an example of a "crazy" state that turned out to be a positive democratic country, in *Cultural Theory* and in *China's Quest for National Identity*. By creating an analogy with Nazi Germany without a clarification of the issues involved in such a comparison the authors leave the parallels to the imagination and emotional response of the reader. Using Dittmer and Kim's definition of a symbol, Nazi Germany used as an analogy becomes a symbol of a "crazy" state, and as a symbol it "expresses and conveys emotion" (p.17). Is the use of this analogy the good value free "science" that Dittmer, Kim, and Wildavsky are so concerned with or does it verge on enemy creation by association?

Filial piety is examined in Chapter 5 of *China's Search for National Identity* by Richard Wilson as an attempt to show a pattern of identity development. Wilson shows how the concept of guanxi (connections) has changed in its cultural meaning from its original Confucianist definition to a contemporary and more encompassing notion of legally enforced equal rights. Wilson establishes that "the way people organize their thought varies with ontogeny, age, and experience .... individual meanings and cultural meanings are not homologous."[49] Cultural meanings for Wilson are embedded in the structure of behavioral rules, and as people interact they interpret the meanings of these rules. Besides the structural nature of cultural meanings, they also serve to coordinate behavior through content.

Content refers to the beliefs and expectations that characterize a society, ... and that frame the ways people are expected to orient themselves one to another. Content is interwoven with particular moral and ethical precepts that justify social stratification in a way that sustains solidarity. Cultural meanings ... have content that defines appropriate orientations toward others through specific injunctions and structure that reveals the sophistication of underlying explanations of social reality.[50]

Using this rule-based definition of meaning (derived from economics) Wilson begins his analysis with the classical Confucianist definition of filial piety and measures this historical definition of social obligation against contemporary legal discussions about social responsibility in the form of rights. His historical analysis of Chinese law shows a cultural shift in meaning from the idea of ones obligations to others to a dualistic, give and take notion of reciprocity. In the final analysis it becomes apparent that there are now regular references to the concept of legal rights which reflects the emergence of "a new sense of national identity,

---

49. Richard W. Wilson, "Change and Continuity in Chinese Cultural Identity: The Filial Ideal and the Transformation of an Ethic," in *China's Quest for National Identity*, 105.

50. Ibid., 106. This is not at all dissimilar to the definitions of group and grid in *Cultural Theory*.

based on the equality of persons combined with communitarian ideals."[51] Yet in spite of this progress Wilson finds it ironic that the Chinese government was able to use the law both to appeal for support and to suppress the students during the Tiananmen Square crisis.

Before providing a possible explanation of this irony I would like to summarize what has been said so far. What have these theories and models tried to uncover through their analysis of international cooperation, domestic political factors, and cultural views? Robert Keohane advocates an international structural analysis to give meaning to state behavior but cannot maintain his theoretical coherency and incorporate domestic factors into his explanations. Stephen Haggard and Beth Simmons emphasize the importance of domestic sociocultural values and the role of domestic learning for giving meaning to state behavior but acknowledge the difficulties of doing this empirically. In looking for domestic indications of democratization and marketization in China, Lieberthal uses implementation to analyze observed bureaucratic behavior patterns; patterns that he interprets as inefficient.

Thompson et al. in *Cultural Theory* also want to explain the meaning behind state behavior. The first sentence of the preface in *Cultural Theory* states, "the subject of this book is meaning." Within the group/grid model, which explains how five different social group types give meaning to themselves and their interactions, a scientific interpretation of these meanings is possible. Lowell Dittmer and Samuel Kim also establish a model for the interpretation of meaning based on the precepts of symbolic interactionism and the "normal" process of development as laid out by psychologist Eric Erikson. This model directs analysis towards the symbols of state power, the myths generated to legitimize current realities, and the change of social realities over time.

All of these theories and models are concerned with the search for an explanation that gives meaning, through scientific

---

51. Ibid., 124.

interpretation, to state behavior. The problem lies in how the concepts of meaning and interpretation are used and not used. The authors that I have reviewed eschew theories that are based on "post hoc observation of values."[52] Thompson et al. say the following in rescuing the concepts of interpretation and meaning from the philosophic school of phenomenology:

> Proponents of hermeneutics, ethnomethodology, critical theory, and the like assert that understanding human beings, because humans confer meaning upon their lives, is inconsistent with theorizing in the spirit of the natural sciences. Our view is that this rigid dichotomy between interpretation of meaning and scientific explanation is unjustified. It is true that human beings create meaning. But it is also true that it is possible to make statements of regularities that help in explaining and even predicting (or retrodicting) the human construction of meaning. Subjectivity need not rule out regularity as long as different sorts of people feel subjective in similar ways with regard to similar objects.[53]

This observation is a response to rescue science's search for an objective reality from the criticisms levied against it by the phenomenological disciplines of hermeneutics, ethnography, and critical theory. It is an acknowledgement of the power of post-modernist theories by inclusion; but if, as the last sentence indicates, science is a majority rule project, even the token acknowledgement of the human construction of meaning is negated. The interpretation of meaning given by a majority is held as true, and their concepts for definition and analysis are not themselves seen as constructed meaning in need of interpretation.

Besides Thompson et al., Robert Keohane also refers to the importance of cultural meaning and interpretation in the analysis of international regimes and cooperation. "Geertz sees culture as the 'webs of significance' that people have created for themselves....It makes little sense to describe naturalistically what goes on at a Balinese cock-fight unless one understands the meaning of the event for

---

52. Keohane, *International Institutions and State Power*, 172.

53. Thompson, Ellis, and Wildavsky, xiii.

Balinese culture."[54] Keohane's points are that cooperation can only be understood within the context of beliefs and expectations and that the study of the functions of international regimes facilitates this context based understanding. Likewise, Erikson's and Piaget's theories of development used in *China's Quest for National Identity* are founded on the principles of symbolic interactionism. Symbolic interactionism is about understanding the multiple meanings that people give to and take from everyday objects and acts. It is based on the understanding that different cultural experiences lead to different interpretations of meaning for the apparently same objects and acts. What is consciously overlooked by these authors in adopting phenomenological principles is that the discussion of meaning has to be a subject-oriented definition of meaning. It is not about meaning as abstracted through a model but meaning as articulated and understood by those peoples and groups being studied. There is a fundamental flaw in wanting to incorporate a philosophy of social reality which is based on trying to understand what acts mean to people within a given cultural context and then to emasculate that same philosophy by forcing an analysis which changes the referent of meaning from the participants to the researcher/ analysis. In this instance the inclusion of postmodern terminology becomes a trapping without its analytical force and undermines rather than enhances the project of social understanding. Two different and equally important lines of questioning are involved here, a post facto search for causality by asking "why" and a present-oriented search for understanding by asking "what is meant."

Understanding and the discussion of meaning are limited by making assumptions about what is known. Asking "what are the variables and the processes of interaction, learning and change?" is a backwards look to explain "why," embedded in an assumption of what goal(s) these actions are being measured against. The hoped for, if not somewhat unrealistic outcome, is that with this question answered social scientists can then plug in objective political

---

54. Keohane, *After Hegemony*, 56.

and economic facts which will result in the understanding, predicting, and controlling of state behavior through economic and political means along a path towards the assumed ideal.[55]

This is evidenced by the concepts which remain unexamined. The chapters in *China's Quest for National Identity* focus on the emerging and existing diversity of groups within China and what that means for centralized political control. In the end it is hypothesized that national or ethnic subcultures may become alternative centers for political loyalty and power. Wilson's specific focus on equal rights is given as another measure of democratization in China. The last chapter mentions strength, compassion, justice, and stability as the "sentimental" concepts which emerged in the book as a "tonal center" that must be played correctly for the appropriate "synthesis" of nation and state.[56] The cultural meaning of these concepts is left assumed and so is defined by the American experience of how these concepts `correctly' blend as virtues to support democracy. Keohane and Lieberthal both rely on the power of market mechanisms in decision making as their ideals of cooperation and decision making among groups. Market mechanisms create incentives and disincentives for certain types of behavior, and decisions are based rationally on an optimum utility function and cost-benefit analysis. Rationality, utility, and cost-benefit analysis are acknowledged to be unique in practice for each different political actor, but in analysis they must be assumed due to the complexity of calculation and for the sake of theoretical parsimony. The result of this assumption is to interpret or define what would be rational or utile from the American perspective.

In the end the ideals of Western liberalism are held as scientific standards for democratization and marketization within China. Although current theories of

---

55. Keohane talks of conditions for encouraging cooperation, Dittmer and Kim talk of the need for positive role models, Lieberthal's conclusion for "appropriate institutions" to minimize or eliminate inefficient bargaining, and Thompson et al. speak of the pluralistic nature of their model and the need to allow pluralism in a society to maintain balance.

56. Dittmer and Kim, 289.

political development and interaction include discussions of change and of the sociocultural aspects of politics, they have not yet escaped Samuel Huntington's charge of "a failure to distinguish between what is modern and what is Western."[57] Making this distinction through self-reflective questioning needs to supplant the idea that the ideals of Western liberalism cannot or will not themselves change over time.

The use of models and theories that maintain unexamined conceptual measures of democracy, rights, cooperation, decision making, pluralism inherently makes an analysis relative to the user's definition of these concepts. Asking the post facto 'why' questions about the structure and nature of relationships is a necessary part of examining the world around us. As an unavoidable human act it establishes categories and regularities that are familiar to people from their own cultural perspective and provides a base for engaging in a dialectical study of politics.

An equally important line of questioning is, "what meanings do the people one studies give to and take from their experiences? What meaning do they give to the concepts of cooperation, democracy, etc.?" By focusing on meaning from this perspective one can see active, culturally informed political acts. This is opposed to 'objective political (f)acts' that can be separated and/or plugged into cultural values with a scientific model.

It is the separation of observed behavior as objective fact from valuative statements that leads to the apparent ironies between what is heard and what is seen.[58] In international relations studies it is the categorization of acts into observable state behavior and unclearly relevant, normative domestic politics that forces a micro-macro dichotomy. It is in this conceptual separation that

---

57. Samuel Huntington, "The Problem of the Study of Political Change," in Louis Cantori, ed., *Comparative Political Systems* (Boston: Holbrook Press, 1974), 400-440.

58. For an excellent study of the fact/value problem see M.E. Hawkesworth, *Theoretical Issues in Policy Analysis* (Albany: SUNY Press, 1988).

researchers fabricate analytic categories and concepts which only have meaning for them and from which they make their own interpretations of the political intentions and expectations of various political actors. The political and historical actions of others are reduced to theoretically convenient supposedly value-free categories.

If the conceptual dichotomy is not maintained or tolerated, there is a greater proclivity to ask for meaning from the participant's perspective because it cannot be accepted or assumed that one's own concepts and categories have a shared meaning, or that they even exist for other people. The conclusions drawn by maintaining that all 'facts' are value laden will be much different from those that are based on an unexamined definition of what is a 'fact,' and yet complementary in that they account for observed behavior and the apparent contradictions of this behavior.

Science is not a majority rule project. It needs to be realized in practice that the answers are based on the questions asked, and that different philosophical orientations raise different questions which are only a threat to those interested in maintaining an existing intellectual or political position. Philosophic complementarity, not hegemony, should propel the investigations of political inquiry. A balance of the opposites through critique and dialogue should be the goal of a mature social science instead of persistently believing that the value of one perspective is inherently the negation of value for the other. The principles of phenomenology and of empiricism may be fundamentally opposed, but the insights they provide are potentially complementary by forcing a more self-reflective, culturally informed analysis of political acts.

The above studies are looking for patterns of democratization. However, if the concept of democracy is explored for culturally different meanings, then the behavior observed and interpreted with these meanings will lead to different understandings of democratization and the possibility of seeing new patterns of political development. The following statement from *China's Quest for National*

Identity expresses the possibility of democracy in China. "Fang Lizhi's stated opinion that the party-state should be based on the support of the people, not the other way around, has found a considerable following, especially among the young."[59] The conclusion drawn from this statement hints that traditional ideas of the relationship between intellectuals and government may be changing to notions of a government based on the support of the general population: the seeds of Western democracy in China.

In contrast, the following illustrates the same situation with a focus on Chinese conceptions of democracy and who the "people" are and what their relationship to the government is.

> Although Western democratic notions are normally linked to pluralism and the free competition of divergent ideas, minzhu (democracy) in China is linked to a vision (Sun Yatsen's) in which people will 'share the same views' and have 'identical' ideals. Nor do the words and deeds of the protestors of 1986-87 or 1989 fit easily with more radical (from a Chinese perspective) Western ideas of direct or participatory democracy. In many cases the students seem to have read the min (people) in minzhu in a limited sense to refer not to the populace at large but mainly or exclusively to the educated elite of which they are a part. This elitist reading of minzhu was clear in the wall posters that appeared in Shanghai in December 1986, many of which took their lead from the speeches Fang Lizhi gave at the city's Tongju University earlier that year. The main theme of these posters, as in many of Fang's lectures and writings, was not that the CCP should be more responsive to the ideas of China's masses but rather that it should allow the intelligentsia a greater voice in national affairs.[60]

Esherick and Wasserstrom write that during the 1989 student movement intellectuals continued to maintain an elitist and inegalitarian definition of

---

59. Merle Goldman, Perry Link, and Su Wei, "China's Intellectuals in the Deng Era," in Lowell Dittmer and Samuel Kim, *China's Quest for National Identity*, 153.

60. J.W. Esherick and J.N. Wasserstrom, "Acting Out Democracy," p.31 in Jeffrey N. Wasserstrom and Elizabeth J. Perry eds.,*Popular Protest and Political Culture in Modern China* (Boulder: Westview Press, 1992).

28

democracy. This discussion of what democracy means to whom by calling into question who exactly the 'people' are and their relationship to the government problematizes Merle Goldman's search for patterns to democratization. However, it is only problematic as long as American definitions of democracy are maintained as the standard for analysis of Chinese democracy.

How does the questioning of concepts in this way apply to Richard Wilson's observation of the apparent irony in the Chinese government's ability to invoke the law, to simultaneously support and suppress the students during the Tiananmen Square crisis. The concepts Wilson uses in his discussion of rights in China must be understood from the Chinese perspective; what is 'equal' and for whom, and what or who is the 'community'? University students that I spoke with in China believed that in 1989 the government acted correctly. The Tiananmen student protestors crossed a political line where they were no longer considered part of the community and so were no longer equal nor deserving of the same rights as members of the community. What is ironic changes when one considers another's cultural base for interpreted meaning. Americans may not like what they see any more or less, but it will not appear to be so paradoxical.

From an international perspective if Western governments believe that the IMF, the World Bank, and GATT symbolize cooperation and movement toward a more stable world economic system, how can this be reconciled with a widely held view in China that these institutions symbolized American hegemony and global economic instability.[61] China uses the World Bank and, by all accounts, is one of the World Bank's best customers, yet if the interpretation of what the World Bank symbolizes is different can Americans assume shared meanings of cooperation or of global stability through economic interdependence?[62]

---

61. David Shambaugh, *Beautiful Imperialist: China Perceives America* (Princeton: Princeton University Press, 1991).

62. China's high status as a customer is based on interviews with Chinese government and World Bank officials.

This section has attempted to show that an analysis of international cooperation must include an active consideration of domestic politics, and that looking at the outcomes of cooperation through policy implementation analysis to understand the relationship between domestic and international factors is a means of doing this. Additionally, I have addressed the concept of meaning and its two philosophically different definitions. These differences are fundamental and should not be confused. Understanding the cultural meanings of those observed can complement through explanation and critique the contradictions of behavior found through post facto searches for patterns and regularities of behavior. This can create different explanations for patterned behavior to be recognized, establishing new sets of concepts and contradictions to be understood.

Conceptual Framework

The interpretive approach to policy analysis maintains that "the belongingness to the world is the interpretive experience itself and that all understanding is mediated by interpretation."[63] The purpose of this approach is to investigate the meanings and values which policy has for the actors involved in a given situation, including the researcher, as a way of understanding the cultural meanings given to policies and social relationships.

"Interpretive approaches to policy analysis introduce a set of questions about how policy meanings are communicated to multiple audiences."[64] These meanings are communicated through organizational artifacts which symbolize

> tacitly known meanings as well as those which are a part of a
> policy's explicit language. Not only do implementors and other
> situational actors interpret these artifacts; the policy and these

---

63. Paul Ricoeur, *Hermeneutics and the Human Sciences*, J. Thompson, Ed. and Trans. (New York: Cambridge University Press, 1981) cited by Marilyn A. Ray "The Richness of Phenomenology: Philosophic, Theoretic, and Methodological Concerns," p. 121, in Janice M. Morse, ed. *Critical Issues in Qualitative Research Methods* (London: Sage Publications, 1994).

64. Yanow, "The Communication of Policy Meanings: Implementation as Interpretation and Text," 41.

interpretations may be 'read' as a 'text' about societal values and identity.[65]

Organizational artifacts are the physical characteristics of an organization, physical in the sense that they can be seen, read, heard, and/or participated in. Artifacts include agency names, program names, procedures, stated goals, reports, etc. "Analytically these artifacts are categorized by such labels as symbols, rituals, ceremonies, stories, and myths."[66] Artifacts are symbolic, or represent relationships; "they stand in as concrete references for more abstract values and beliefs and are shorthand ways of communicating those meanings."[67] Artifacts can be objects, such as emblems or statues. Language is also an artifact in its reflection of values and the multiple meanings that it can convey. Actions such as rituals and ceremonies become symbolic artifacts in that they are the enactments of the values and beliefs which they represent. Stories narrate events which represent values of importance, and even the act of telling the story are symbolic. In this study stories include professional research papers as well as verbal stories.

Stories emerge that attempt to explain what is seen but not fully understood. Dvora Yanow refers to these forms of story telling as myths. Myths "reconcile two or more incommensurable truths; they end further inquiry, and convey a sense of something of value."[68] Myths evolve naturally, they are believed, they are contemporary, and they are often seen in problem-setting stories. Policies can support multiple and conflicting goals that are exposed by examining the stories which explain

---

65. Ibid. For examples of interpretive policy analysis see also Janice Love and Peter C. Sederberg, "Euphony and Cacophony in Policy Implementation: SCF and the Somali Refugee Problem," *Policy Studies Review* (Autumn 1987): 155-173; and Robert T. Nakamura, "The Japan External Trade Organization and Import Promotion: A Case Study in the Implementation of Symbolic Policy Goals," in Palumbo and Calista, *Implementation and the Policy Process: Opening Up the Black Box*, (Westport: Greenwood Press, Inc., 1990): 67-86.

66. Ibid., 47.

67. Ibid.

68. Ibid.

In summary, organizational and policy artifacts can be conceptualized as symbolic objects, symbolic language, and symbolic acts.

> Artifacts are 'read' in a particular context: they carry the meanings of a particular point in time, or of a particular socio-cultural environment. This means that artifacts accommodate multiple meanings. Meaning is not universal or determinate; it depends on context and on the perception and interpretation of the participant. When meaning is shared, the artifacts create a feeling of unity among those who share the same or similar interpretations of them and demarcate those people from others who hold different interpretations. Since artifacts accommodate multiple meanings, it may not be self-evident that two parties do not share similar interpretations.[69]

In this book I will examine the meanings of the title "Chinese Ecological Agriculture," as a metaphor, the implementation of the policy as a ritual, and the policy as a myth that supports the conflicting but interlocked realities of government and farmer goals. As will be seen, the interpretive analysis of this policy reveals what I believe to be an important aspect of Chinese political culture, a symbolic reality that I refer to as model following. After showing that the policy of ecological agriculture is symbolic of ideal social values beyond the specific scope of agricultural development, I bring the key concept of model following into the American academic dialogue about Chinese international behavior. In doing this I want to provide an alternative, Chinese-based interpretation that can account for the inconsistencies left by American explanations and one that encourages a critical self-reflection of the primary concepts widely used for international relations analysis.

It is through the example of my alternative explanation that I intend to illustrate a different perspective on the levels-of-analysis issue. I believe that the problem for analyzing the linkages between domestic and international events is one of interpreted meaning. I propose that by using domestically informed

---

69. Ibid.

32

constructs in international analysis, recognizing current mainstream theories as a specifically domestic American construct, I can redefine these levels as a problem of interpreted meaning rather than strictly as a problem of conceptualizing processes and practice. Using domestic politics to learn about cultural identities and political cues that interpret meaning to inform international analysis bridges the level-of-analysis gap conceptually rather than treating the researcher and the theoretical concepts as removed from practice. This is an important shift because it is these concepts that define observed processes and shape their interpretation and the resulting analysis.

My goals are to apply the interpretive method to the specific domestic policy of agricultural development as a means of (1) learning about Chinese domestic politics and political culture; (2) applying what I learn about a Chinese political identity as an interpretive framework that has implications for American domestic and international analysis of China; (3) and, lastly, advancing the idea that international processes, policies, and institutions need to be conceptualized symbolically as artifacts that are inherently interpreted rather than as objective conditions to be explained.

The remainder of this book is divided into five additional chapters that will detail the process of how I learned about an important aspect of Chinese political identity and how I apply this learning to the analysis of Chinese domestic and international behavior. Chapter 2 provides a discussion of the analytic concepts, the methodology used to collect the information for this book and the problems, situations, and cultural assumptions I encountered. In Chapter 3 I discuss the history of agricultural development in China since the 1978 reforms, the context for the emergence of Chinese Ecological Agriculture, and a detailed explanation of the techniques of ecological agriculture. The examination of the history and the practice of this project gives a preliminary picture of the nature of political relations and the uses of power in Chinese domestic politics. Chapter 4 will be an analysis of the policy of ecological agriculture and its implementation from an interpretive perspective. From this analysis the concept of model following emerges and it is a symbolic reality that manifests perceptions of power and social structure. Chapter 5 will focus on the Chinese literature of ecological agriculture as symbolic of an ideal social structure and values that go beyond agricultural development in their political and cultural significance. The analysis presented here will be used in a comparative way to examine some of the American literature on Chinese political identity. The last chapter, Chapter 6 will show how the theme of model following, as the major cultural theme to emerge from my analysis, provides an alternative to neorealist explanations of Chinese international cooperative behavior. In this chapter the hermeneutical circle is completed, because that which is learned through the analysis is appropriated as a domestically grounded conceptual framework that can be used as an alternative explanation for international analysis and understanding.

CHAPTER 2

**LEARNING TO LISTEN: A METHODOLOGY FOR THE
ANALYSIS OF METAPHORS, RITUALS, AND MYTHS**

The purpose of this chapter is to establish and define the concepts and methods used in my research and analysis. First, I will define in more detail the concepts of metaphor, ritual, and myth as used in my analysis, and second I will describe my research experience, the collection of materials in China, their analysis, my expectations and the realities of my research encounters. My method of language analysis involves three moves. In the process of interviewing I 'unpacked' key concepts introduced by speakers during conversation. By 'unpacking' them I examined the meaning of the concepts in the context of the speaker's dialogue by asking for more information or explanation of those ideas. In doing this I came to focus on key metaphors that were used for explanation and clarification of concepts to me, in particular the concept of ecological agriculture as a the program title and as a metaphor. Next, by focusing on the implementation of the policy as a ritual act, I discovered that the concept of using and creating models is an important ritual and a significant cultural theme. Last, I found that the contradiction between the high possibility for large scale adaptation of the project and the continuation of government development and extraction policies which limit those possibilities could only be explained by considering the policy in its entirety as a myth. This is an application of the concept of myth, not as a created false consciousness but as a naturally occurring means of explaining and reconciling contradictory realities.

Symbols, Meaning, and Interpretation

Symbols arise from the need to "summarize and index knowledge and experience." Particularly in novel and unfamiliar situations, they arise from the need for communication with common reference points for categorizing shared experiences among a people and for distinguishing among people and establishing or affirming social identities.[1] In short all social activities rely on the use of symbols to simplify, clarify, and give meaning to ones social experiences. They are "used for any object, act, event, quality, or relation which serves as a vehicle for a conception - the conception is the symbol's 'meaning.'"[2] The concept is different from the object it is attached to; it is the difference between the physically observable, such as a building or words, and the concepts which give meaning to the observable, for example the building design or the idea of democracy.

Only man among living things reconstructs his past, perceives his present condition, and anticipates his future through symbols that abstract, screen, condense, distort, displace, and even create what the senses bring to his attention. The ability to manipulate sense perceptions symbolically permits complex reasoning and planning and consequent efficacious action. It also facilitates firm attachments to illusions, misperceptions, and myths and consequent misguided or self-defeating action.[3]

---

1. Charles Elder and Roger Cobb, *The Political Uses of Symbols* (New York: Longman Press, 1983), 31. For a philosophical discussion of hermeneutic thought see Josef Bleicher, *Contemporary Hermeneutics: Hermeneutics as Method, Philosophy and Critique* (London: Routledge and Kegen Paul, 1980); Paul Ricouer, *Hermeneutics and the Human Sciences: Essays on Language, Action and Interpretation*, ed., trans. John B. Thompson (Cambridge: Cambridge University Press, 1981); Paul Ricouer, *Interpretation Theory: Discourse and the Surplus of Meaning* (Ft.Worth: The Texas Christian University Press, 1976); Michael J. Shapiro, *Language and Political Understanding: The Politics of Discursive Practices* (New Haven: Yale University Press, 1981).

2. Clifford Geertz, *The Interpretation of Cultures* (New York: Basic Books, 1973), 91.

3. Murray Edelman, *The Symbolic Uses of Politics* (Urbana: University of Illinois Press, 1964), 2.

The interpretation of experience, of language and acts, is a process of giving meaning to those experiences by organizing new information to fit existing cognitive structures. A recent conversation illustrates this point. A tourist visiting the fishing community where I now live inquired about the type of fishing that was taking place here. "It depends on the season, but now it's for salmon, rock fish, and cod," I replied. Still he seemed unsatisfied with the answer and asked, "Aren't they fishing awfully close together?" It occurred to me then that he was specifically asking about the boats in the small harbor, and when asked about this he confirmed my suspicions. He thought it peculiar that all of the boats were gathered so close together in the harbor to fish such a small area. I explained that these boats were not fishing but were simply moored in the harbor waiting to be used for fishing at sea. Although his observation seemed odd to me, namely, in his failure to note a lack of people or fishing equipment on the boats, I must surmise that in his experiences boats not in use had to be in slips, and boats out of slips must be in a state of use, especially in a fishing village. He had a pre-existing conception of how boats in port should look, and since these boats in this harbor did not fit that cognitive structure they must be engaged in a different activity; the boats and not the cognitive structure were questioned.

How people understand or interpret political acts is no different; they are interpreted and given meaning by existing cognitive structures. There are several features of political acts that make them distinctive. Political acts are about power and the exertion of influence over the distribution of resources, and as such the stakes are higher for those wanting influence and power. The meaning or perceived meaning of these acts tends to be "mutually reinforced in large collectives of people, evoking intense hopes and fears, threats, and reassurances."[4] Therefore, understanding meanings, their creation, and their transformation as a function of symbolic apprehension is necessary for explaining political acts.[5]

---

4. Murray Edelman, *Politics as Symbolic Action: Mass Arousal and Quiescence* (Chicago: Markham Pub. Co., 1971), 2.

5. Ibid.

Political power and action is more about changing or maintaining demands and expectations by using or generating social cues than about the tangible outcomes of policy. Social cues are political acts that trigger meanings regarding group status and security and their future status and security. Political acts and policy proclamations are signals and assurances that specific group interests will be considered or ignored, and they influence the perception of potential adversaries and their possible behaviors. Group expectations on the other hand influence what is perceived to be a relevant fact, and the interpretation of those 'facts' that results in specific attitudes towards policy acts and signals.[6]

Because policies derive their meaning from the social cues as well as from the effects they have through their implementation, it becomes apparent that a single act can cue multiple meanings. A policy that protects the interests and/or status of one group can be viewed as a threat by another. The abortion issue, medical reform, and continued political support for Taiwan are examples of policy issues that have cued different symbolic meanings for different social groups both in the language used and in the acts carried out. Broadly speaking, the abortion issue symbolizes different concepts of individual legal rights and social obligations, whether priority goes to the mother or the fetus, and how those rights are expressed legally and morally. Changing the allocation and monitoring of medical benefits symbolizes different conceptions about the role and limits of government and private businesses to supervise and make decisions about access to and forms of medical attention offered. Continued support in the United States for Taiwan symbolizes a distrust of Communist China and different understandings about the idea and role of a single China as the basis for good economic and political relations.

Actions, verbal and physical, themselves grounded in pre-existing, culturally defined rules, cue existing cognitive expectations so that the perception and interpretation of the 'facts' by groups supports or challenges their position

---

6. Ibid., 8-10.

relative to others. The recent visit by the president of Taiwan to Cornell University as an act was the visit of a graduate to his alma mater. The act became symbolic because of the different interpretations given to it. For Communist China the visit symbolized a violation of the concept of a single China ruled from Beijing, and for the pro-Taiwan group of Americans who invited him, the visit symbolized the need to support a 'democratic' China and an outpost against Communist China. Both interpretations of the visit were derived from and served to support existing beliefs about the role of Taiwan in international politics.

If the conventional study of politics is about the struggle for power and the allocation of benefits, how people get what they want, then the study of symbolic politics is about how speech and non-speech acts influence what people want, fear, believe possible, and sense as their identity and community. The study of symbolic politics involves a redefining of leadership and therefore power in terms of the ability to shape impressions and build coalitions of support rather than trying to define ideal types of leadership or measure the results of a policy against the stated intentions.

Symbolic politics is the politics of using culturally recognized symbols to reassure a community and in doing so to build up the power base of leaders responsible to or at least in need of a given community's support. The roles that social groups take will determine the interpretation of and the subsequent response to those policies. The patterns of social stratification, the degree of urbanization and industrialization, and the definitions of acceptable behavior place severe limits on the number and types of policies that will be popular or acceptable. In a dynamic and changing social environment people's perceptions and attitudes are fixed only as long as identify with the same social role and the symbols that are effective for that role. Once roles change so do the perceptions, attitudes, and expectations that influence the interpretation of meanings.[7]

---

7. Ibid., 186.

40

By examining political acts as symbolic one can recognize that they will cue different interpretations to different people. This acknowledgement forces us to look beyond the empirical data that provide only one perspective in a multifaceted world. To bring multiple perspectives together for comparison and contrast is not an attempt to reduce two views into one but is an effort to understand the social structures and cues that support multiple meanings; structures that facilitate change and stability in role identity and the interpretation of political acts.[8]

Metaphors

Metaphors are a figure of speech in which a word or phrase literally denoting one kind of object or idea is used in place of another to suggest a likeness or analogy between them, for example, drowning in money.[9] Metaphors are not simply rhetorical or literary devices that can be added or removed without changing the meaning of a text; they are ever present in daily conversation, in all levels and forms of discourse.

> Our ordinary conceptual system in terms of which we both think and act is fundamentally metaphorical in nature. Our concepts structure how we perceive, how we get around in the world, and how we relate to other people. Our conceptual system thus plays a central role in defining our everyday realities.[10]

The metaphors of 'war is hell,' or 'time is money,' or 'heaven is high and the emperor is far away' are examples of how metaphors shape one's ideas about war, time, and local behavior through a culturally specific understanding about

---

8. Edelman, *Politics and Symbolic Action*, 7.

9. *Webster's Ninth New Collegiate Dictionary* (Springfield, Mass: Merriam Webster, Inc., 1984), 746.

10. George Lakoff and Mark Johnson, *Metaphors We Live By* (Chicago: University of Illinois Press, 1980), 3. For a current overview of the role of metaphors see Dvora Yanow, "Supermarkets and Culture Clash: The Epistemological Role of Metaphors in Administrative Practice," in *The American Review of Public Administration* 22, no. 2 (June 1992): 89-110, and also Andrew Ortony, ed., *Metaphor and Thought* (Cambridge: Cambridge University Press, 1993).

hell, money, and a distant, ultimate authority. The understandings of the concepts are grounded in the lived experience through our bodies (motor skills, perception, mental capacity, emotion), our physical environment (movement, manipulation), and with others within our culture (social, political, economic, religious). Descriptions of our experiences are culturally specific, and this specificity determines the concepts that are used, how they are used, and the meanings that are given to them.[11]

Lakoff and Johnson outline five characteristics of metaphor that shape the meanings people give to their experiences.[12] I would like to build upon their example of the metaphor 'time is money' to illustrate the characteristics of a metaphor. The basic definition of a metaphor is a comparison of two terms with different literal meanings, which invites a comparison and creates a defining image of one in terms of the other. Time is a culturally defined measure of the interval between events or conditions. Time may be measured in relationship to earth seasons, to the rising and setting of the sun or moon; it also may be measured in life spans as in generations, to growing seasons, the interval it takes for sand to fall from the top to the bottom of a container, etc. Society has divided the passing of time into years, months, weeks, days, hours, minutes, seconds, and smaller. With this division one can measure the passing of time and the time it takes to do something or to wait for something to happen. Money is a unit of exchange or a measure of value. Money can be shells, salt, gold, eggs, paper, animals, or any number of items that have an agreed to value, usually based upon scarcity and therefore desirability of the medium. Time is not a medium of exchange, nor is money a measure of time. Labor that takes a certain amount of time or saves time may be an acceptable exchange, services for goods perhaps,

---

11. Ibid., 56.

12. Lakoff and Johnson believe that one of the most dramatically forceful metaphors in terms of its global impact by facilitating the spread of modern capitalism, is the metaphor, 'time is money.' I have chosen their example, and examine it in greater detail as a way of illustrating the characteristics of metaphor.

but I cannot give someone five minutes as a separate item. It is only by describing time in terms of money, coined by Benjamin Franklin, that time can take on the qualities of money as a unit of value and of exchange. By combining two concepts that have different literal meanings, the reader is provoked into rethinking the nature of time.

Metaphors highlight certain features and suppress others. To say that time is money focuses attention on efficiency, work, and the urgent need to save and earn capital; it says nothing about the value of time spent in non-work-related activities. Additionally, it elevates the concept of time to the status of an object, a resource that must be managed and not wasted. The emphasis is shifted from a focus on the work itself to one that is mediated by time and what is considered to be 'good' work as measured against time as a subjectively defined standard.

Metaphors entail not just other concepts but very specific aspects of those concepts. To continue with the time as money metaphor, time is not historical time or geological time or daytime/nighttime but is specifically referring to chronographic time that is spent on the concept of work. It more specifically refers to the ability to manage the efficiency of work as measured against the goal of using less time, conjuring up a specifically capitalist notion of time: capitalist because it is the capitalist who loses the most from inefficiency and capitalist because money is the object obtained. The metaphor is not 'time is a paycheck,' 'time is a pension,' or 'time is happiness,' but time is money, and it creates the image of unlimited potential to get money if time is managed efficiently and primarily engaged in work.

This begins to illustrate the third point, that by highlighting and masking experiences, metaphors give concepts new meanings. The focus on chronographic time diminishes the importance of seasonal time or of daytime/nighttime or of biological time. Modern technologies have at once expanded the concept of time by making physical time irrelevant and by shortening the amount of time it takes to do work. Work is conducted around the clock, around the world more quickly than ever before.

Metaphors can be appropriate because they sanction actions, justify inferences, and help set goals. The new president of a company I worked for decided that too much time was spent in the break room between 8:00 and 8:30 in the mornings and that too many people were arriving late, after 8:00 a.m. He began taking down the names of people who arrived late in the mornings to be announced publicly in weekly division head meetings, and he also made a point to end the morning discussions in the break room. From his point of view time was being wasted and therefore the money that the company could earn. His goal was to increase efficiency and profits, and this sanctioned the actions that he took. In doing this he did overlook the point that much of the talk in the break room was frequently about work-related issues at a time of day that was not yet interrupted with meetings and incessant phone calls and in a location that allowed interdepartmental communication to occur in an informal setting. Further, although he was concerned that the salaried employees arrive at 8:00 a.m., it was not a problem that often they stayed until 7:00 or 8:00 p.m. to finish their work.

Lastly, the meaning of a metaphor will be partially cultural and partially tied to personal past experiences. 'Time is money' defines how Americans work in modern culture, and the advent of newer and faster technologies has compressed the amount of time needed to produce goods, which at once saves money by decreasing costs and increasing production, and both increase profits. My earning of a Ph.D. has generated a number of comments over the years which I believe informally reflect the metaphor at a personal level. When I first made the decision, I heard comments such as, "Why are you leaving such a good job?" or "You won't earn any money with a Ph.D." More frequently I have been told that with a Ph.D. I will be earning a lot more money. In one instance when I gave my more personal reasons for wanting a Ph.D. I was not believed, the assumed 'real' reason being that I wanted to earn more money. In this case the metaphor of a 'job is money' was more defining than perhaps the metaphor of a 'job is satisfaction.' My decision to spend time on a Ph.D. was seen either as a detractor or as a boon to my earning power, but in either case money was the reason and the objective, not more personal or idealistic motivations.

> New metaphors have the power to create a new reality. If a new metaphor enters the conceptual system that we base our actions on, it will alter that conceptual system and the perceptions and actions that the system gives rise to. Much cultural change arises from the introduction of new metaphorical concepts and the loss of old ones.[13]

In this study 'ecological agriculture' is treated as a metaphor. As the name of an agricultural development program, it is the combining of the concept of agriculture with the concept of ecology that forms a new metaphor which defines the goals, actions, and assumptions of the program. As a metaphor it highlights the concept of agriculture as understood from the Chinese experience and with the addition of the concept of ecological, which in its most simple sense refers to the interrelatedness of all living things with each other and their physical environment. This modification of agriculture as ecological creates a new meaning which builds from past experiences, sanctions specific actions, and through its specific focus on method and resource integration suppresses attention to the continued lack of government investment in agriculture.

Ritual

> Metaphors provide organizing images which ritual action puts into effect. This ritualization enables people participating in ritual to undergo apt integrations and transformations in their experience. The study of ritual is the study of the structure of associations brought into play by metaphoric predications upon people.[14]

---

13. Lakoff and Johnson, 145.

14. James W. Fernandez, "The Performance of Ritual Metaphors," in J. David Sapir and J. Christopher Crocker, eds., *The Social Use of Metaphor: Essays on the Anthropology of Rhetoric* (Pittsburgh: University of Pennsylvania Press, 1977), 101. I have substituted the word 'people' where the author uses the word 'pronoun' to refer to people.

Ritual is the action that involves the individual in social activity and that gives meaning to the concepts which are symbolized by the metaphors, myths, and objects. It is the point of interaction that is observable and often accessible to the nonparticipant observer.

In this study the concept of implementation was chosen as the enactment of policy that would most likely provide an opportunity to observe and discuss the nature of the interaction between different groups involved in the project. The forms of interaction would be the same forms that would reflect separate perceptions about the nature of the policy and that would change the policy in its actualization by compromising and bending the policy to fit with their understanding of it. It was this understanding of policy implementation that was the basis for many of my assumptions about what I would find, and one that was quickly dispelled.

I found that implementing the ecological agriculture project is not a conflictual undertaking, nor is it the forced imposition of a government decreed policy. Instead its implementation was based on voluntary participation and the rituals of repetition and of following models. It is by repeating, individually and through the mass media, the principles and methods of ecological agriculture and evidence of their effectiveness established in government supported models that the idea of ecological agriculture is spread. Establishing models and the idea of having a model to follow are important rituals that define acceptable actions and motivate by establishing a goal achieved through individual self-improvement. As will be more fully explained in Chapter 4, both of these rituals have cultural significance that extends beyond the reach of communist influences. They are ritual that reinforce ideal forms of social relations and self-identities.

These rituals actualize the metaphor 'ecological agriculture' making it an individual experience that connects the individual to society and reaffirms the social experience by associating it with the individual. Through the enactment of these rituals the metaphor achieves a fundamental transformation on one hand through the socialization of individual experience and on the other hand by

46

rendering the social experience individually relevant. "Ritual is motor activity that involves its participants symbolically in a common enterprise, calling their attention to their relatedness and joint interests in a compelling way that promotes conformity and evokes satisfaction."[15]

Myths

> Myths present culturally defined truths in the form of stories, parables, and aphorisms that simplify, highlight or dramatize basic cultural premises and prescriptions. They offer socially constructed accounts of exemplary behavior and significant events in the life of the polity. Their primary purpose is to instruct.[16]

By providing instruction and a way to understand actions and events myths can at least temporarily hold competing values in tension by rationalizing anxiety and directing attention elsewhere.

There is the presumption by many that myth is "associated with the primitive, the past, the subjective or the false. Such judgements assume that modernity itself, or some isolated and sophisticated part of it, has outdistanced myth and can speak from a privileged, myth-free stand point."[17] This however is not the case, myths are an integral part of the social landscape that help us to understand and explain that which is contradictory or confusing by directing our attention away from points of tension. Another way to understand this

---

15. Edelman, *The Symbolic Uses of Politics*, 16.

16. Elder and Cobb, *The Political Uses of Symbols*, 54. It is important to add that the concept of culture as I would like it to be understood applies to separate social groups, to organizational cultures, and to a larger idea of a society. Myths appear in all of these settings and their functioning is the same. For an overview of the literature on myth see Murray Edelman, *Politics As Symbolic Action: Mass Arousal and Quiescence* (Chicago: Markham Pub. Co.,1971); Dvora Yanow, "Silences in Public Policy Discourse: Organizational and Public Policy Myths," in *Journal of Public Administration Research and Theory* 2, no.4, (October 1992):399-423; Anthony Cohen, *The Symbolic Construction of Community* (London: Tavistock Publications, 1985); Milton Scarborough, *Myth and Modernity* (Albany: SUNY Press, 1994); Alan M. Olson, *Myth Symbol and Reality* (Notre Dame: University of Notre Dame Press, 1980).

17. Scarborough, 13.

contemporary notion of myth would be as a rationalization of confusing situations that meets the exigencies of a given point in time.

Operating with incomplete information and interpreting what is perceived through what is expected, stories and myths emerge that explain, justify, and prescribe action. One creates real understandings that highlight what one wants to believe and overlook or downplay those elements which confuse the understanding of a given situation. It is the degree to which a myth or a rationalization's central premise fits with people's existing expectations that the myth has power.

Dvora Yanow outlines four basic elements of a myth: narration, social construction, belief, and incommensurability.[18] Myths are matter-of-fact statements that are resistant to factual attack. In the case of China's ecological agriculture, it is the myth of possibility, enacted in the ritual of model following. This myth implies that it is very possible for every individual by following the model and creatively adapting it to local conditions, that he or she can benefit economically, participate in the process of China's modernization, and still work within the contemporary and historical constraints of high population density and limited resources. The possibility for individual action and the potential for success were immune to factual attack in the sense that it could not disproved or seen as impossible or as weaknesses to the program when the political and economic problems of implementation were discussed. In the end, I was told, with patience and creativity of application, people can make it work.

The great focus on the power of the individual farmer in the myth of possibility, which is certainly a significant factor, also turned attention way from larger economic and political difficulties that affected individual implementation, not the least of which is increased government extraction of resources from rural areas. For individual farmers the lack of local political support needed to get permission to try to implement ecological agriculture, or the lack of money to

---

18. Yanow, "Silences in Public Policy Discourse: Organizational and Policy Myths," 401-402.

purchase equipment and animals necessary to diversify the farm's production, essential to the implementation of the method, are practical issues that undermine the acclaim to possibility.

Myths are produced in response to the needs of the moment. They evolve to explain contradictory realities and are not conscious lies or intentionally deceptive fabrications. Ecological agriculture as a concept has been emerging in China since the mid-1970s. Its acceptance as policy in 1986 came at a time when agricultural output began to decline after the initial gains from the 1978 reforms and when government investments in industry increasingly came to be at the expense of agriculture. It is a program of agricultural production that emerged as independent research, motivated by Western research on sustainable agriculture, and one that has real environmental and economic benefits for both farmers and the government.

Myths are believed because they allow one to know; they express what exists. Ecological agriculture addresses China's contemporary economic and social problems in matter of fact terms that are supported by individual experiences. As a myth it allows farmers to discuss social and economic problems as historical and national problems. It places the individual farmer into a larger context that minimizes apparent disorder and inequalities and provides a means for the individual to take positive action that will benefit the household and the nation.

Myth is exposed when the point(s) of tension or contradiction is openly discussed. One of the purposes of this study is to examine ecological agriculture not only as a technical program but in its relationship to China's development goals. The point of tension for ecological agriculture is revealed with an examination of what is required to implement the method and which farms have been most successful. Implementation requires a degree of capital investment that is unavailable to many farmers and belies the easy adaptation the discussion of the method implies. Capital investment primarily comes from government sources, which is minimal, or from farmers already wealthy from industry. The rural/urban income disparities, the minimal government investment in agriculture, and the need to stabilize population movements to urban areas are the very points of social tension that ecological agriculture overlooks explicitly, but implies in its promise of increased income and its requirement for increased labor.

Symbolic politics and analysis involve understanding the multiple meanings that concepts have for different groups and how political acts cue the interpretation of meanings that generate support for both change and continuity. Metaphors are the expression of one concept through the combination of two seemingly disparate things or ideas that create an image or feeling in the mind of the interpreter. They are fundamental to the shaping of ideas and the categorization of experience. In shaping ideas metaphors create identities which authorize specific actions. Myths are stories or explanations that emerge over time to explain contradictory realities, and incommensurable policy goals. Myths are believed because they express and explain a reality for those who believe it. Rituals are the physical actions which reinforce the tenants of the myth. It is through ritual that the concepts contained in the metaphor and myth are enacted to make them socially relevant to the individual and in turn revitalizes the social power of the myth and the metaphor. How these concepts were employed in my analysis of Chinese ecological agriculture will be explained in the next section.

Interpretation and Analysis

I spent the fall of 1994 at Nanjing Agricultural University, Nanjing, Jiangsu Provence, in the People's Republic of China. During that time I was able to interview thirty people including farmers, hamlet, village and township leaders, researchers, and government officials of Jiangsu Province in a combined total of 38 different interviews. I was able to speak with leaders of a state farm, of a village with ecological agriculture, and a hamlet without. I had the unexpected surprise of an interview with a township extension worker from Anhui province over an extended lunch at the China Environmental Protection Agency office in Nanjing. He was there only for the day, and agreed to speak with me about his work. I was able to meet several times with Jiangsu Province Department of Agriculture officials and extension workers at the provincial level in Nanjing, in both formal and informal settings. At the county level I had the opportunity to meet with extension workers from one county with and from one without the project. I met with researchers and extension workers at the National Environmental Protection Agency of China, at the Jiangsu Academy of Agricultural Sciences, and with agronomists and agricultural economists at Nanjing Agricultural University. I also arranged meetings with faculty members of the Department of Sociology at Nanjing University and the Shanghai Academy of Social Sciences towards the end of my trip to discuss my understanding of model following in Chinese society. My interviews also included Westerners (four Americans and an Australian) who had lived in China for a number of years. Two of these interviews were specifically about agricultural development; parts of these interviews and others were about Western observations and impressions of Chinese culture. My interviews included a number of high-level administrators and officials as well as interviews with lower-echelon staff members. Among those I interviewed were the director of Technical Development at the National Environmental Protection Agency, the director of the National Society of Sociology, former director of the Agriculture Modernization Institute, and the directors of Ecological Agriculture at DaFeng and SiYang counties and officials from the Jiangsu Province Agriculture and Forestry Department.

My interviews were one and a half to two-hour interviews which resulted in seventy-six hours of taped and written interview material. Eight of my interviews were repeat interviews that allowed for a more in-depth exploration of material from the previous interview and the introduction of information from other interviews as a way of balancing view points and perspectives. A number of my interviews, especially in the counties and my initial Ministry of Agriculture interviews, were group interviews with up to six people present. In these settings it was interesting to note the acknowledgement of hierarchy and authority. Often question answering was deferred to the CCP representative, or highest authority.

Responses by other members had to be specifically elicited by me, or permission delegated/granted by the group leader.

Some of the interviews were conducted in English, but the majority were in Chinese. Although I do speak Chinese, I used an interpreter during the interviews to ensure understanding and maintain the flow of conversation. I tape recorded many of the conversations, but I also found that unrecorded interviews were much more relaxed and I believe more informative. There was often a hesitation, especially by those closest to the bottom of the hierarchical ladder, to be recorded at all. In some cases they refused, and in other instances when it was explained that their supervisor had been recorded, then they agreed as well. The recording created a situation of formality, and I believe that it was the permanence of words on tape and the possibility of saying the 'wrong' thing (which I inferred from the hierarchical and deferential nature of the group meetings to be a real possibility) that made these recorded sessions more official. In the unrecorded interviews there was a more interactive, back and forth form of conversation and a greater easiness in answering my questions that probed for the clarification of a story or an idea. A preliminary analysis of emerging themes and issues was done while I was in China to help me in conducting my interviews and maintaining a programmatic perspective. In addition to the interview material I collected approximately fifty articles (some in Chinese and others already translated into English) which include technical writings about ecological agriculture (from individual researchers and the Chinese journal, *The Journal of Ecological Economics*) and conference papers from the 1994 Conference on Integrated Resource Management in Beijing and the 1995 Conference on Sustainable Agriculture also Beijing. These articles provided me with information about Chinese sustainable agriculture, and its role in Chinese economic and social development. I used these articles to analyze which topics were discussed, how they were discussed, and what values were reflected in the language used. I also maintained a thorough journal of my daily activities, cultural self-reflections, impressions, and feelings.

52

The purpose of my research was to examine the implementation of ecological agriculture as a contemporary policy issue that may add to our understanding of changes and continuity in Chinese political culture. My hope was to clarify the meanings of concepts used by different groups involved in the project's implementation. By looking for points of conflict and convergence in these meanings, I wanted to learn about what the Chinese understand themselves to be doing and what significance it has for them and for non-Chinese observers. In this way, this analysis of ecological agriculture serves as a case study to understanding meanings which are cultural issues and are relevant for domestic politics and international behavior.

There are several areas of significance relevant to my research methods: (1) the assumptions of what I expected to find, (2) the reality of what I found and how I found it, and (3) my analysis of what I found. I believe that the best way to air these issues is as a brief narrative of my experiences and thoughts roughly categorized into the three areas of my assumptions, my research encounters, and my analysis. A discussion of my experiences is important not only so that others can understand what I did, but also so that I can illustrate the self-reflective nature of hermeneutic research and the importance of listening self-reflectively to what is being said, of listening from within.

Learning to listen from within involves learning to listen to what other people believe is important for one to know, what they want to tell one, and learning to listen to one's own inner voice as it reacts to and interprets what is being heard and seen. I came to learn that invariably the opportunity would arise within each interview to ask the specific questions that I was interested in if I was patient and listened to what I was being told. If I began with questions that were not from the context of the conversation, it was clear in the answer that I received that the meaning I had intended was not what had been understood. Listening to what the speaker was telling me provided me with new information and a new perspective, and it also created a context within the conversation for asking my own specific questions either to clarify a concept or to introduce a new point for discussion.

This realization became especially clear in an interview with a senior agronomist at the Jiangsu Academy of Agricultural Sciences. Prior to the interview I had started to realize the importance of setting up and using model farms as a means of implementation, and I wanted to learn more about the concept of model farms. I began my interview by asking if he would tell me about the nature of his work and specifically his work with ecological agriculture. This repeatedly proved to be a good way to put people at ease and to find out their own perspectives on the project. After a lengthy discussion of tillage management methods and the problems of adapting no-till methods to his research site, I was wondering how I could use what was being said to direct the conversation back to the model farms.

He kept speaking of his methods of no-till as 'new technology,' and I asked him, "How well do the farmers accept this new technology?" Now the interview became more interesting for me too; I was listening to him discuss the problems of farmer's educational and experiential levels and the additional problem of the high cost of technology. From this point I asked about the problem of cost and where farmers on model farms got the needed monies. I was told about national funding for himself and demonstration fees to local leaders, as well as local government support for new tools and machines. This discussion about money was not a comfortable one for my interviewee; his knees were wiggling back and forth at an increasing pace until he ended the money portion of the conversation by saying that the costs for no-till are relatively low because they use few tools and his research is low cost because it relies on comparisons of different demonstration sites over the period of several years. From this I asked how what he learned was passed on to the farmers. The benefit of the demonstration sites was in providing the farmers and village leaders a chance to see for themselves the benefits of new methods and to decide for themselves whether or not to adopt these new technologies. For me this had been the first real articulation of how model farms were supposed to work and generated many questions that I used in future interviews.

Throughout the whole conversation I was listening for key words that would serve as natural lead-ins for my own questions. For me this was a process of learning new terminology and using these terms to ask what I wanted to know. At the same time I was working with my informants to keep them a part of the interview dialogue and not as subjects of an interrogation. When I began the interview I had not anticipated asking this person about money, yet that portion of the conversation was one of the most informative and came up because I was listening to what was being said. Sources and uses of money is also a touchy subject for most people anywhere and had I begun with such an intense question or asked it at an inappropriate time I would not have received the detailed answer that I did. In the end I came away armed with new terms and insights about the relationship between farmers and researchers and the role of models in passing on new technology.

In addition to listening to others I had to learn to listen to myself. Listening to my own inner voice was a process of becoming aware of and questioning of my own emotional responses and intuitions about my experiences and about the information I was hearing.[19] My emotions ranged from curiosity and bewilderment, to frustration, anger, and disgust, and as time progressed to a sense of perspective, understanding, and growth. I had to focus constantly on the information I was hearing and seeing to clarify points that I did not understand for both technical and cultural reasons. What were my expectations as an American about the type of answers I would hear; how did my social position as an outsider affect what I was told. I became aware that my reactions of distrust were based upon notions of interaction with a communist regime that I did not realize were so

---

19. Sherryl Kleinman and Martha A. Copp, *Emotions and Fieldwork*, Qualitative Research Methods Series 28 (London: Sage Publications,1993). This monograph provides a discussion of the emotions and feelings that go with field research and techniques for incorporating those feelings into the research/learning process.

much a part of my own conceptual make-up. In all of these cases I had to take my emotions and interpretations to task and ask myself, "Why am I feeling this way, or why am I making these conclusions? What part is the result of culture clash and what part can I trust as being a valid response to the situation." The notion of validity was particularly tricky, because all of my responses were valid, but it was a matter of sorting out the issues that evoked a given response. The product of this soul searching was to come up with new questions and directions for interviewing and to gain a cultural perspective that included myself in the research process.

In preparation for conducting interviews, prior to my trip, I had generated a list of possible questions to ask of the different groups of people with whom I would talk.[20] I had categorized the groups into government officials, researchers, and farmers. Each group had questions that corresponded to their areas of expertise and responsibility. The questions covered general areas of technology, economics, and implementation. Two general questions that I had determined as important to ask everyone however where, "what are the goals of ecological agriculture," and "how do you think other people (i.e. those not in the group I was interviewing) view the project?" The purpose in asking these questions was to begin to establish a point of comparison for seeing how different groups viewed the project and where potential points of conflict and cooperation were in the process of implementation.

---

20. The following books provided a basis for the conduct of my research and bibliographic resources: William Foote Whyte, *Learning From the Field: A Guide From Experience* (London: Sage Publications, 1984); Peter J. Frost, et al, eds., *Reframing Organizational Culture* (London: Sage Publications, 1991); Janice M. Morse, ed., *Critical Issues in Qualitative Research Methods* (London: Sage Publications, 1994); Michael I. Harrison, *Diagnosing Organizations: Methods, Models, and Processes* (London: Sage Publications, 1987); Dvora Yanow, "Reflecting on Methods: Garbage-can Cases and Footholds of the Mind," Chapter 2 of *How Does a Policy Mean* (unpublished manuscript, 1994); Harry F. Wolcott, *Transforming Qualitative Data: Description, Analysis, and Interpretation* (London: Sage Publications, 1994); James P. Spradley, *The Ethnographic Interview* (New York: Harcourt Brace Jovanovich College Publishers, 1979); Robert K. Yin, *Case Study Research: Design and Methods* (London: Sage Publications, 1989).

This rational was based on my assumption that implementation was and is a conflictual process. I believed that this would also be true in China where more American research is contradicting the assumption of absolute state power in forcing uniform change by showing the influence of local and regional officials on national policy and a high level of conflict among various groups.[21] I also assumed that I would in some degree become aware of these conflicts through the different views that each group would have of the project, in terms of its relative benefit and perceived problems. My knowledge of ecological agriculture prior to my trip consisted of a general understanding of the concept and related agricultural techniques. My primary focus in preparing for my research involved clarifying and operationalizing the conceptual categories that were to guide my analysis.

The first of my interviews were with a chief agronomist at Nanjing Agricultural University who is also one of the leading figures in the development of ecological agriculture. My relationship with him from the beginning was always tutorial. I sensed that he viewed me as the naive, but eager freshman and he was the senior professor who would entertain my questions and enjoy watching me learn. This tutorial sense accompanied me in different degrees throughout nearly all of my interviews and was as unsettling as it was productive.

My job as a researcher was to examine concepts and ideas that arose in conversation; to unpack their meaning as much as possible in order to facilitate cross-cultural understanding on my part. The people I interviewed often expected me to be a technical expert in the field of agriculture or economics, and though I asked pertinent questions I felt that they were often seen as too basic, lacking technical depth. However, the questions were only basic if I too accepted the idea that I should be a technical expert.

---

21. For a good example of this research see: Kenneth Lieberthal and David Lampton, *Bureaucracy, Politics, and Decision Making in Post-Mao China* (Berkeley: University of California Press, 1992).

I was frequently reminding myself that my technical expertise was in political analysis not in agricultural methods, that I was not exclusively interested in the technical issues but in how they were discussed. I believe that not only was I dealing with cultural differences in Chinese/American terms, but I was also dealing with differences in professional cultures. My interest in having concepts explained, by frequently asking, "What do you mean by ..." or "Why is that important?" helped to cross these cultural barriers and to keep people interested in talking to me. The tension for me was in living two cultures (an American in China, and a political scientist in agronomy) removed from my own experiences and to be continually searching for reference points of understanding. This desire for common understanding especially noticeable when someone would make an analogy that was also a political joke; I had to forgo the shared laugh, and instead ask for it all to be explained.

Towards the end of one two-hour interview when I was told that Chinese policy making is like "a moonlight policy," I just wanted to relax and enjoy a mutually shared joke. Instead I had a rather blank look on my face and wondered if this was a reference to the need for people to hold second jobs. How far this was from understanding the intended meaning was soon revealed. "Moonlight policy," I was told, means that the policy changes constantly like the phases of the moon.

I used semi-structured interviewing throughout my research period, that is, to have an open-ended type interview that I would guide with more specific questions. My first lesson was that my original list of questions made no sense at all to those I interviewed. My concepts and categories were either seen as missing more important issues or else had no meaning at all. When I asked about how new technologies were introduced and adopted by the farmers I received a rather quizzical look and was told that everything is initiated by the government and the farmers do what they are told as long as it works. This answer was of interest, but not at all what I was really trying to learn about.

As I listened to the story of ecological agriculture I found out about the importance of using models for implementing the project. When I realized the importance of the concept of the model and then rephrased my questions in this new context I received much more informative answers. My new questions were more along the lines of the following: "What does it mean that ecological agriculture is a model of agriculture?" "Why is having a model farm effective?" "How are models selected?" "What is a successful model?" My original question about the introduction of new technology was answered, but only after listening and learning about and using the concepts that had meaning for those I spoke with. The effectiveness of my listening and questioning improved significantly when I could see my original and subsequent lists of questions being answered from the Chinese point of view without having asked the questions in their preinterview form.

Everyone that I spoke with had their own agenda, what they felt was important to discuss. Usually it was about themselves and their work. I learned that instead of asking questions that were of interest to me about the project, I needed to listen to what the interviewees were telling me that they thought was important. In asking questions about what I had just heard (clarification of concepts, why the issues were important, etc.), I learned about the project and gleaned information and insights that not only answered my unasked research questions but also generated comments I might not otherwise have received, and I maintained participant interest throughout the interviews.

When I was in the position of directing the interview with my questions I was focusing only upon what I felt was important, and this was without exception not what was important to the other person. All interviewees had their own take on the project and their own responsibilities that they identified with as being particularly significant. When I acknowledge their views and priorities, I was acknowledging them as important in the process of developing or implementing the project. By allowing people to talk about their interests relevant to ecological agriculture, I found their interest in talking to me was high, and when I introduced a question that was of interest to me within the context of their own interests I received answers that were informative and given in an atmosphere of teaching or helping me.

Before each interview I listed the questions or issue areas that I was concerned with at the top of my note page. As the interview progressed I listened for concepts that were significant and made notes to ask what was meant. It was in the discussions of clarification that I would find or create a context for my specific questions of interest. Although each interview was offered new information and questions particular to that conversation, it also was an opportunity for me to ask questions about my growing understanding of ecological agriculture and emerging themes and issues.

Two general sets of questions became useful to me, those relating to concepts used in the discussion of the project and those about problems or difficulties in implementing the project. Some examples of concepts that emerged in conversations were: 'Chinese Characteristics,' 'models,' ecological' as a defining concept of agriculture, 'good social values,' 'government responsibility for macroplanning,' and 'the dragon's head' to list a few. Asking about these concepts as tangents to issues raised by the interviewee proved a good way to shift naturally, even if temporarily, the questions to concepts that I wanted clarified and to corroborate answers among interviewees.

Questions about difficulties were treated matter-of-factly, and I allowed some time to be spent discussing practical issues that were of specific concern to the interviewee. Two main themes emerged: lack of money and older leaders not open to new ideas. The second point was mentioned quickly and lightly, and it was clear that I should not press my questions in that direction. Through my discussions about money, or lack of money, I became aware of the tensions or contradictions in implementing the project. As I learned about the difficulties in getting the money necessary for initial implementation and that the more successful farms had government funding or already rich owners I began to think more critically about the project both at the local level and as a national policy. I began to ask slightly different questions; some were very pointed, such as "It seems that the more successful model farms have nonagriculture sources of funding. How are less fortunate farms supported in following the models?" The answer to this question, which will be discussed more fully in Chapter 4, was that it was their problem and the current focus of work was on improving the model of ecological agriculture to be followed. This emphasized to me the importance of two emerging themes: the importance of creating, being, and following good models and the idea that the approaches to problem solving in China which I saw are not what I am used to as an American.

After every interview I wrote a summary of my notes and questions for the next day. I also kept a journal in which I recorded my impressions, feelings and ideas. It was in this journal that my most critical and reflective questioning occurred, particularly in trying to understand my feelings about what was happening during the interview process. I believe that strong feelings of confusion, anger, or frustration reflect points of cross-cultural differences that need to be understood self-reflectively and not immediately used to interpret and categorize the actions of others. As mentioned earlier, it was this process of listening from within that gave me the greatest opportunity for learning about the project of ecological agriculture, my own professional and social cultural predispositions, and the relationship between my own concepts and how they affected my interpretation of experience.

My strongest feeling was the result of my initial contacts with the official line about ecological agriculture. The fact that I was so consistently being told the same thing by a variety of people created a strong sense of frustration and the impression that somehow I was being lied to. I questioned why I felt so strongly that I was being lied to and had to consider other reasons for the behavior I was observing. First were my own preconceptions about the nature of an authoritarian, communist regime that penetrates all levels of society and generates official 'lines' to explain, motivate, and educate people in the 'correct' way. My readings about Chinese politics and the nature of party meetings to establish the party line had affected my beliefs in a way that I had not anticipated, and which is linked to my second set of assumptions.   My second set of assumptions were based upon my being an outsider, that is non-Chinese. I found out later that being an outsider in China is a concept of exclusion not limited to non-Chinese people. It begins at the family and village level and reflected an unanticipated cultural bias that expects greater homogeneity at a level that does not exist.

I expected that I would hear a variety of views openly expressed. This is what I would encounter in a group interview or individual interview in the United States, a general agreement about the principles of a topic that would quickly degenerate into each individual's take on the goals, problems, and solutions. I anticipated a much more opinionated dialogue that would more immediately open itself up to directed questions. When I was faced with such consistent responses from my Chinese hosts, there were the immediate and now obvious clash with my assumption of open opinion and the feeling of uniformity and collusion that emphasized to me my position as a cultural outsider. My understandings and expectations of degrees of openness were confused and contradictory. I expected more diversity of opinion and found greater unity, though the diversity became apparent after I worked through this initial confusion learning to listen and ask questions in a different way.

At the same time I had to trust that my instincts were not completely invalid. This bracketing of my own intuition as culturally invalid helped me to accept what I heard as reflecting at least a partially if not wholly believed reality, and an appreciation for what I believed helped me to maintain a critical awareness. In this way I learned to ask different questions that would either passively clarify or critically extend my understanding. It was the feeling of frustration and my self-reflection that helped me to place what I was hearing and seeing into a larger conceptual and analytical context, that forced me to change my interview techniques, that allowed me to see the official response as a part of the policy process (as a ritual behavior), and that opened my own expectations to inquiry.

62

In my visits with hamlet, village, and township leaders, party officials, and Department of Agriculture officials, the hierarchy of authority and protocol was obvious, and subordinates were unwilling to say or do anything that would be seen as wrong or improper by their supervisors. Often if there was an uncertainty about allowing the interview to be recorded or about the appropriateness of answering a question, my interpreter would tell them, truthfully, that one of their senior officials had done it and then they would also agree. Only twice was I asked not to record an interview, and all of my questions were answered. Yet, I could not seem to get past what quickly became apparent as an 'official response.' The response was familiar in its terminology used to describe the problems and goals of ecological agriculture: increasing population, decreasing resources, the need for more food, and the holistic philosophy behind integrated agricultural techniques. Initially all of my questions elicited the same response, and different tacks on the same question only initiated the same response from a different departure point. Once I began listening and asking question from the context of what I had heard rather than persist in using my original list of questions, then I moved past the repetitiveness of the official answer.[22]

I had two components to my analysis, one was the analysis of my interview data and the other was the analysis of the written materials that I had acquired. Ever present were the questions, "what does the information I am getting mean" or "how does it relate to the analytical concepts that I brought with me?" I think these questions loomed large because of the intense relearning that was going on during the process of my interviews and the realization that none of my assumptions were working here and that I was having to question myself as well as the interviewees.

---

22. There were areas that I didn't have access to, for example my questions about the farmers and my access to rural areas was limited and controlled. The farmers I did speak with gave what would be appropriate answers for discussions with an unfamiliar outsider, that is they clung closely to the official line. People who had been farmers and were now in government and party positions were more forth coming. The limitation of access was not detrimental to my overall research for I was able to find answers to all of my questions during my relatively short stay. Any limitations could only have been overcome with a much longer stay in China of one or more years.

Though I had made interim observations, or analysis of what I was hearing and seeing, it was not until the last weeks of my stay that I was able to put the information into a more coherent set of ideas. Frequent rereading of the articles and notes that were influential in defining my conceptual framework (outlined in the first half of this chapter) served as a reference point and guide for interpreting the information and concepts that I was learning about. This was particularly true in dealing with the increasingly large contradiction between the high social and environmental value of the project and its relatively limited and difficult implementation process. It was only after this tension reach a critical mass in my own mind from mounting evidence that I realized that this was the contact point between two conflicting realities that allowed me to discuss the policy as a myth.

I had anticipated that my analysis of the written materials that I acquired in China would help to provide corroborating support for my analysis of the interviews. What I did not expect was the degree to which they would allow me to see ecological agriculture in a larger sociopolitical context. Following the structuralist interpretation of narratives set out by Eloise Buker in *Politics Through a Looking Glass: Understanding Political Culture Through Structuralist Interpretations of Narratives*, I treated each professional article as a story narrative.[23]

I first would learn who the antagonist and protagonist of each story were in order to establish the opposites that were reflected, i.e., Chinese/Americans, researchers/peasants, chemical fertilizer producer/ organic fertilizer proponent. A second reading was a careful search for metaphors used to describe the relationship between the opposites described above. Lastly, I sought to understand what transformations were expected or had occurred in order to understand the values that are sought through change and who or what served as the agent of change. By looking at the desired transformation one can learn about

---

23. Eloise Buker, *Politics Through a Looking Glass: Understanding Political Cultures Through Structuralist Interpretations of Narratives* (Westport, Conn.: Greenwood Press, 1987).

desired values, by examining tensions one learns about conflicting values, and by learning about characters one learns about who or what is considered to be an effective agent of change.

Several important points emerged, the details of which will be discussed in Chapter 5. My initial expectation was to learn more about ecological agriculture specifically and about societal relationships in China; I was surprised to see that the primary conflicts were between Western and traditional Chinese methods, adopting modern science without loosing a Chinese identity. It became apparent to me that when looked at as a whole, the ecological agriculture project papers became a metaphor for an ideal social structure, not just an approach to farming.

Ecological agriculture is presented as a model for integrating Western science with Chinese values and as a model for an ideal, integrated, sustainable social structure. Also of note is the way the Chinese talked about themselves and how Americans, and some Europeans, talked about the Chinese. The Chinese see themselves more as our near equal, wanting to learn from our experiences and to eventually become the ones from whom Americans learn. The Western perspective is that the Chinese are regionally strong but still underdeveloped. They are in need of American help and should follow its advice to overcome significant economic and social obstacles. The difference in attitudes and interpretations of the current situation is significant for understanding identities and sources of current and potential conflict. The result of my analysis of written material was to be able to see ecological agriculture in a cultural and political context that went beyond the confines of only a technical program for agricultural development. As a symbol for an ideal integration of resources, ecological agriculture represents the social, political, and economic structures necessary to ensure a holistic, balanced, and sustainable growth.

Interpretive policy analysis is about understanding ideas and actions from the perspective of the individuals one is learning about. In this study the policy artifacts that became most significant were the program name Ecological Agriculture as a metaphor for understanding and action, the use of models and repetition as ritual behaviors which activated the metaphor and implemented the policy, and the program goals as two sets of valid but conflicting priorities supported by the same policy when it is conceptualized as a myth. As a policy ecological agriculture relies upon cueing cultural interpretations of meanings which are widespread and significant enough to motivate action. These cues are not generated artificially but are a generally acceptable set of actions that educate and motivate individual action. Understanding these cues and the meanings that are given to them is an examination of the "symbolic forms through which each community tells itself and other audiences something about its identity as a polity."[24]

---

24. Dvora Yanow, "The Communication of Policy Meanings: Implementation as Interpretation and Text," in *Policy Sciences* (Spring 1994): 57.

CHAPTER 3

THE CONTEXT AND METHOD FOR ECOLOGICAL AGRICULTURE

The purpose of this chapter is to provide a brief overview of the reforms that took place in the period of 1978 to 1986 and then to examine in detail the methods of ecological agriculture. I have focused on this eight-year period to highlight the political, social, and economic changes that were taking place in the years leading up to 1986 when ecological agriculture was officially adopted and supported as a national policy. In doing this I hope to establish a context for understanding the policy that incorporates the political and developmental considerations of the national leaders and their relationship with the peasant population. The second section of this chapter will detail the techniques and principles of the project as a method of agriculture. This section will provide clarification into the practice of ecological agriculture, complementing the historical description of power plays with one vision for achieving integration and harmony.

This historical survey will be a descriptive narrative of the political conditions that preceded the emergence of ecological agriculture. I will exclude analyses that focus upon the implications or effectiveness of specific policy decisions relative to democratization or marketization in order to focus more closely on the political interaction between China's peasants and the government in developing and implementing agricultural policy. The purpose for doing this is to begin to explore issues of identity and power that are not considered in studies that interpret power through the lens of democratization and marketization. It is with this historical perspective that I want to lay the ground work for my analysis

68

in the remaining chapters, a foundation that illustrates why I believe in the need to consider explanations that are based more fully on Chinese understandings and uses of power.

I have found that local studies of the reform period have proved most useful for offering a distinctly Chinese-based understanding of the politics of reform. These works are also beneficial in that they go beyond a perspective that deals primarily with the authority and power of the state to a perspective that considers the interactive nature of the peasant/state relationship. I have found David Kelliher's book, *Peasant Power in China*, to be very illuminating in this regard and have relied on it heavily for the historical and political discussion in this section.[1]

Political and Economic Reform

The goal of Chinese agricultural policy is to achieve self-sufficiency in food by the year 2000. China maintains 21% of the world's population (1.3-1.5 billion), which is increasing by approximately 12-17 million people a year, on 7% of the world's arable land. Domestically, the resource constraints are equally

---

1. Prasenjit Duara, *Culture, Power, and the State: Rural North China 1900-1942* (Stanford: Stanford University Press, 1988); William Hinton, *The Great Reversal: The Privatization of China, 1978-1989* (New York: Monthly Review Press, 1990); Philip C.C. Huang, *The Peasant Family and Rural Development in the Yangzi Delta: 1350-1988* (Stanford: Stanford University Press, 1990); Daniel Kelliher, *Peasant Power in China: The Era of Rural Reform 1978-1989* (New Haven: Yale University Press, 1992); Sulamith Heins Potter and Jack M. Potter, *China's Peasants: The Anthropology of a Revolution* (Cambridge: Cambridge University Press, 1990); Helen F. Siu, *Agents and Victims in South China: Accomplices in Rural Revolution* (New Haven: Yale University Press, 1989). There is a tendency for Western analysts to overlook or understate Chinese concepts of power and social relations. As the following quote illustrates there is often the frustrated feeling by Western observers that if only the government did X, then Y would be develop smoothly. This does not consider that for the Chinese government the current relationship between X and Y is one of power and control, and that 'efficient' political or economic growth is entwined in what is viewed as necessary for maintaining social stability in conjunction with economic growth. "So long as prices and incentives are closely linked with government finances, budgetary considerations rather than the aim of promoting efficient agricultural growth will continue to interfere with the making of farm policy." Terry Sicular,"China's Agricultural Policy During the Reform Period," in Joint Economic Committee, Congress of the United States, eds., *China's Economic Dilemmas in the 1990's: The Problems of Reforms, Modernization, and Interdependence,* (London: M.E. Sharpe, 1992), 364.

significant; only 25% of China is arable land, and it has only one-fourth of the world's fresh water reserves. The problems underlying the goal of self-sufficiency are the need not only for increased production but increased production within the constraints of limited and diminishing resources.

Government functions in agricultural development have shifted since 1979 from a direct micromanagement involvement in "entrepreneurial functions" in the commune system to a more administratively removed role of "making public policies to influence farmers' behavior."[2] This shift allows greater individual influences on implementation, which also increases the impact of individual self-interest on achievement of the larger goals. There are three main areas of policy change that have occurred since the end of the collective system of farming: (1) the adoption of the household production responsibility system (land reform), (2) an increase in the procurement price of agricultural goods (price reform), and (3) privatization to foster village and township enterprises, which has led to the emergence of labor and credit markets.

These policy changes resulted in a rapid diversification and growth of agriculture. However, a number of consequences have further complicated the initial problems that reform was intended to address: (1) grain production has not continued to grow consistently at expected levels,[3] (2) the implementation of these structural changes has not necessarily affected a lasting change in agricultural methods,[4] and (3) resource depletion from these farming techniques

------

2. Zhu Ling, "The Transformation of the Operating Mechanisms in Chinese Agriculture" *The Journal of Development Studies* 26, no.2 (January 1990), 231.

3. Joseph R.Goldberg, "Grain Options for China:1990-2000," in T.C. Tso, *Agricultural Reform in China*, 114-122.

4. The Chinese press mentions problems of technology absorption and problems in rural areas with an understanding of markets. The more pressing problems of resource scarcity is also a recurrent theme. Traditional farming in China is family based, utilizes small plots of land which are often not adjacent to one another, use unsophisticated equipment and antiquated land use and management techniques, i.e. slash and burn methods, and soil exhaustion, and maintains sustenance production levels. "Perfecting the Household Contract System," FBIS China (Dec. 4, 1990), 45; "Water Conservancy" FBIS China (Dec.4, 1990), 36; "Agricultural Circular issued by CPC Central Committee and State Council," FBIS China (Dec. 7, 1990), 36; "Vice Premier Tian

and from nonagricultural usage is an increasingly significant problem.[5]   To consider these problems and the conditions that gave rise to ecological agriculture I will examine the following four policy areas: (1) family farming, (2) price reform, (3)privatization, (4) credit and labor. Through these vignettes patterns of interaction, identity, and power emerge that are relevant as parts of the model following concept that emerges at the end of Chapter 4. It is from this picture of struggle that the policy of ecological agriculture emerges as one possible means of achieving greater political harmony and peasant inclusion in the process of economic development.

Family Farming

In 1978 China's weak agricultural sector was the main obstacle to achieving the goals of the Four Modernizations - science and technology, industry, agriculture, and armed forces - by the year 2000. China's leaders recognize that strong agricultural production is fundamental to achieving the other three goals, not only in terms of food and resources but also for the money that it provides for industrial and urban growth. "The development of industrialization under socialism demands that agriculture supply industry with ever increasing quantities of commodity grain and industrial raw materials and that it accumulate funds for construction and provide an expanding market."[6] Between 1953 and

---

Jiyun speaks on stabilization of contract responsibility system," FBIS China (Nov. 15, 1991), 54. A survey of Chinese farmers indicates that the majority are still working at subsistence farming for themselves, in *Beijing Review* (Sept.7-13, 1992), 18.

5. See T.C.Tso, *Agricultural Reform and Development in China*; He Bochuan, *China on the Edge: The Crisis of Ecology and Development* (San Francisco: China Books & Periodicals, Inc., 1991); Cheng Xu, Han Chungru, and Donald Taylor, "Sustainable Agricultural Development in China," in *World Development* 20, no. 8 (August 1992), 1127-1143; Piers Blaikei & Harold Brookfield, *Land Degradation and Society* (New York: Routledge, 1987).

6. Wang Songpei and Zhu Tiezhen,"Lun woguo nongye jitihua de guanghui daolu" (The Glorious Path of Our Nations's Collectivization of Agriculture), *Jingji Yanjiu (Economic Research)* 6 (1981), 45, in Kelliher, 46.

1978 the urban population had increased by 120%; however in 1978 the net grain production was only 19% greater than in 1953.[7]

In 1979 Deng's reform coalition proposed that the reforms include higher procurement prices paid by the state, a small expansion of private plots within a predominately collective land system, incentives to increase work by increasing pay for more work, and a 'responsibility system.' The responsibility system was call this by the Chinese leaders because it was a plan that guaranteed a contractually specified level of remuneration for collectives living up to their 'responsibility' in achieving state quotas.[8] The Central Committee directives did not dismantle the collective system but encouraged greater responsibility by smaller teams of workers. The problem for Chinese farmers lay in how not to work as a fully integrated collective and yet remain structured as a collective. According to Kelliher it is the confusion created by unclear directives from the party leadership about how to implement the 'responsibility system' that led to the innovative interpretation by farmers and collective leaders to establish household contracts between families and the collectives. The result, arrived at through experimentation, was a reorganization of farming to a level of individual responsibility that was traditionally acceptable and economically beneficial. As a result of the unofficial spread of household contracting throughout China, the Central Committee found itself in a position of playing policy catch-up. In 1978 household contracts were illegal, but over several years they came to be viewed by the state as first, not desirable, and then finally in 1981 permitted as a national policy.

Household contracting was a significant departure from the intense level of state control expressed through the collective system. In 1979 Zhao Ziyang, the Party Secretary and later Premier of Sichuan Province advocated the 'Four Don'ts' as a way of passively acknowledging the reality of household contracting

---

7. *Statistical Yearbook of China*, English Edition (Oxford: Oxford University Press,1985), 185.

8. Kelliher, 56.

and its challenge to the collective system.[9] The 'Four Don'ts,' which quickly spread throughout China as guidelines for dealing with peasant innovations were don't publicize, don't oppose, don't support, don't stop. Zhao realized that open support would bring in strong political opposition to the radical changes taking place; however he believed that by "letting many different creations develop freely meant that only an innovation capable of succeeding by its own tenacious merit would survive. Once a successful innovation became a fact of rural life, it would be too late for opponents to stop it."[10] The attitude of allowing peasant innovation spread and in 1981 was supported by Deng Xiaoping with the following statement: "We should let every family and every household think up its own methods of doing things, and let them find more ways to raise production and increase income."[11]

There were several reasons for the rapid spread of the household responsibility system. The emergence of family farming in the 1978 reform period was the third time that family contracting, contacting with the government to be responsible for the production on a specific piece of land, had become a means of production since the revolution. From 1955 to 1957 family contracting was allowed in order to maintain the familiar landlord-tenant relationship as the state moved to consolidate power in the rural areas with cooperatives as a first step to collectivization. From 1959 to 1961 household contracting emerged again, with state support, to keep production going during the disaster of the Great Leap Forward. The continuity and familiarity of household contracting were never really lost.

After the failure of the Great Leap Forward and the chaos of the Cultural Revolution, the peasants exhaustion and distrust of the government were at an all

---

9. Zhao Ziyang rose as an ally of Deng's reform movement to become the Chinese Communist Party Secretary-General. He was replaced by Jiang Zemin in the aftermath of the 1989 student demonstrations due to his failure to maintain 'order.'

10. Kelliher, 79.

11. Quoted in *Renmin Ribao (People's Daily)* (May 20, 1981), 4, in Kelliher, 79.

time high. Peasant dissatisfaction came to be seen by China's leaders as a potential threat to the success of the reforms. The peasants were not a direct threat to state power, but their ability to affect the production of crops through passive means such as lying, reporting false harvest figures, foot dragging conniving against outside officials was not underestimated.[12] The state leaders realized that they could not rely on coercion to force compliance. They needed to gain the peasants more willing cooperation. It was pointed out by Chen Yun, the reform coalition's economic chief, that "if the peasant's income is raised, the state stands to benefit the most," monetarily and politically.[13] At the Third Plenum it was declared that the "first task of modernization was to 'release the socialist enthusiasm' of peasants, to 'pay full attention to their material well-being.'"[14] Family farming through the responsibility system was the means to this end.

Working in the commune system did diminish personal incentives to a degree, but this is not the only reason for their abandonment. Collective farms were not a universal failure, but the central government's penchant for nationally uniform policies prevented different regions from practicing what worked best for them. William Hinton emphasizes this point in his book *The Great Reversal: The Privatization of China, 1978-1989*, stating that many of the collectives were able to produce a surplus or at least satisfactorily meet their own food, and income needs. He goes further by saying that presently in China some areas still practice collective agriculture as the more efficient means of production. They are taking an inventive step now in the same way that family contracting was creatively initiated in 1979 by the farmers. In my own interviews in China, I was told that the more avant guard research is focused on reaggregating land for large scale,

---

12. Kelliher, 53.

13. Chen Yun, "Chen Yun's Speech at the CCP Central Committee Work Conference," translated in *Issues and Studies* 16 (April 1980), 95, in Kelliher, 55.

14. Du Runsheng, "Zai Zhongquo nongcun fazhan wenti yanjiu zu taolunhui shang de jianghua" (Speech at the Conference of the Research Group on Problems of Chinese Rural Development), *Nongye jingji congkan (Agricultural Economic Perspective)* 3, (1981), 2, in Kelliher, 54.

mechanized production, which is an important step in the re-popularization of an old idea.[15]   The scientific research by think tanks and the academic community gives justification to government policies as scientifically base and socially beneficial.   This work not only advances the work of improving agricultural techniques but also serves to calm fears and distrust that is a part of the contemporary peasant attitude towards government enforced changed.

Decollectivization gave a degree of control back to the farmers.   The owning of land has historical and economic appeal by virtue of the status and security it provides.   Homes and land are inheritable property (homes legally; land is a de facto inheritance).   Land also provides security against the uncertainty of future policy changes, and the ability of one to always have land to farm for food in troubled times.   Land is divided for inheritance to sons, and it is divided within villages to maintain an egalitarian redistribution in the face of loses to industry and urban development.   General Secretary Hu Yaobang noted that "Dividing the fields is a habitual idea among the peasants,"[16] and the degree to which fields are divided is extraordinary.

It was explained to me by an environmental researcher that in addition to family reallocations to children, every time farmland is lost to housing or industrial development the remaining farmland is reapportioned to maintain egalitarian standards within a community.   The result is that the small plot of land that a farmer is cultivating this year may or may not be his the following year, and the plots that do belong to a single family are not adjacent, but are scattered throughout the village area.   Furthermore, the crops grown by the village are also

---

15. William Hinton, *The Great Reversal: The Privatization of China, 1978-1989* (New York: Monthly Review Press, 1990).
Author interview, October 1985, Nanjing, People's Republic of China. The general distrust of government and the fear of recollectivization is also discussed in David Zweig, "Peasants and Politics," in *World Policy Journal* 64, no. 4 (Fall 1989), 643.

16. Hu Yaobang, "Zhongyang shujichu tongzhi tingqu nongcun gongzuo huiyi huibao de chahua" (Remarks by Comrades of the Central Committee Secretariat while Listening to the Report from the Conference on Rural Work), reprinted in *Zhongqong yanjiu (Communist Party Research)* 16, no. 4 (April 1982), 107, in Kelliher, 96.

apportioned so that each farmer is responsible for an equal percentage of each crop. This means that each farmer in a village will be growing a little bit of everything in disconnected plots.[17] The problem is twofold. First the growing of crops is very inefficient since a lot of people are growing little bits of the same crop instead of promoting efficiency by growing larger amounts of a single crop in the same area. Second, farmers who take care of their land one year may lose it the next in a reapportionment and gain land that has not been well cared for, reducing any incentives to put too much work into properly fertilizing and working the soil.

Family farming in the now popular form of the Household Responsibility System was the result of farmer initiated interpretations of party policy guidelines. The state did want to increase farmer's enthusiasm for production in order to increase material and monetary resources for industrial development but did not necessarily want to give up the social control that the collective offered them. China's leaders could have enforced collectivization against the widely popular family farming, but they feared the economic and social cost of alienating such a large portion of the population. This should not be seen as a unique form of policy formation, as David Kelliher points out, "deriving policy from local experiment is a common theme in the history of the People's Republic of China."[18] This in a direct contrast to the common Western notions of Chinese policy making as solely a function of high level government policy creation and implementation through an undemocratic process. Instead this reflects a policy process whereby achieving a goal such as agricultural development is the priority, and the process used to achieve that goal is not a formal process but is one that by virtue of interpretation is more flexible and open to local adaptation than is usually expected by outside observers.

---

17. Author interview, November 1994, Nanjing, People's Republic of China.

18. Kelliher, 70.

Price Reform

The procurement system in China is a system designed to secure the peasant's surplus production for the state, to extract capital from the countryside, and to control peasant cropping decisions.[19] Prior to 1979 there were three different state procurement prices: (1) there was the list price for the basic quota, (2) there was a premium price for extra-quota sales (which were mandatory quotas for grain, cotton, and oil crops), and (3) a floating price for voluntary sales beyond the first two quota prices. In 1979 the list price and the premium prices paid by the state were raised on all grain and oil crops. The premium price was raised from 30 to 50% above the list price, and after fulfilling the first two quotas a farmer could sell the remaining surplus on the free market or to the state at a negotiated price.

The peasants quickly learned a variety of ways to manipulate the pricing system by selling at the premium and negotiated prices in order to earn the maximum profit from the state. One way was to ignore the quotas for all crops other than the state procured products and in this way grow enough to sell in the extra-quota and negotiated categories. Another method was to hold back on storable crops, such as corn, for several years. The real output would be hidden, and the officials were told that there had been a poor harvest. Then in a single year, as a coordinated effort, all of the saved corn, being enough to meet the higher quota levels, would be sold at premium prices. Production teams and

---

19. Philip Huang in *The Peasant Family and Rural Development in the Yangzi Delta* elaborates on this process of state involution has having deep historical roots, and Prasenjit Duara also discusses state involution as a part of the Chinese development experience. Involution is the process of state extraction of capital from local areas, which foments increased production, then increased extraction from fixed resources. It is a cycle that historically is unsustainable and results in economic and political collapse, though Huang argues for the possibility of involution with economic growth through diversification.

Duara, in *Culture, Power, and the State* makes the following comment about involution, "The increased demands of the state led to the proliferation of entrepreneurial brokerage, and this proliferation led to yet higher demands." The relationship between the entrepreneur, which Duara describes as a local bully the modern equivalent of imperial state brokers, creates a relationship between the state and the broker that resists change, and has a "profoundly disintigrative and delegitimating impact" (Duara, 251-57).

families would cooperate to make a variation on this method. One team would claim a poor harvest and give all of its grain to the cooperating team. The second team would combine the harvests of both teams together, sell it all to the state for the premium price, and then split the income with the first team. Another, more confrontational means would be to refuse to sell to state procurement agents at the list price. The procurement agents, under deadline and quota pressures from their superiors, would give in and sell at higher negotiated prices, at a cost to the state, rather than face the angry peasants.

In all three cases the greater the supply of grain the greater was the cost to the state for procuring the established quotas. State officials began to realize that something was amiss when the records were indicating that in spite of declining basic quota requirements there were still constant shortfalls, and yet the total procurement levels continued to increase as did the state's expenditures. In 1985 Central Document Number 1 announced new, apparently liberalizing, policies to prevent the farmers from continuing to defraud the government.[20] The government eliminated mandatory quotas, multi-tiered pricing, negotiated procurement, and production assignments. Now there was only 'contract purchasing.' State grain and cotton companies were to negotiate purchasing contracts before each season. Any surplus goods after the initial contract requirements were met would be bought by the state at the old list price.

The purpose of price reform was not economic liberalization, but to "eliminate the policy of premium prices and wide open procurements, and adopt a policy of suppressing prices."[21] In achieving these aims the policy was a success. In 1986 prices were 10% lower for grain and 13% lower for oil-producing crops. Grain growers lost money in their sales to the state while market prices went up

---

20. "Zhonggong zhongyang, guowuyuan guanyu jinyibu huoyue nongcun jingji de shi xiang zhengce" (Ten Policies of the Chinese Communist Central Committee and the State Council for Further Enlivening the Rural Economy), from *Central Document No. 1 for 1985*, reprinted in *Renmin Ribao (People's Daily)* (March 25, 1985), 1, cited in Kelliher, 136.

21. Gao Hongbin et al., "Chang xian zengzhang, yihuo fazhan chizhi" (Normal growth or sluggish development), *Jingji Yanjiu (Economic Research)* 9 (1987), 49, in Kelliher, 136.

78

29% in 1986 alone due to poor harvests. The state had saved money and through pricing was again in control of the peasant's fortunes.[22] It was in 1985 that the fortunes of the reforms up to that time began to reverse.

The conservatives, led by Chen Yun, won out in their concern for maintaining state control rather than allowing increased peasant autonomy in pricing and marketing decisions.[23] In addition to the issue of state control there was also a national security concern that rising farm prices would destabilize industrial development and the state budget. There was and is a three-way relationship between the state, industry, and agriculture. The state depends upon industry more than agriculture for its financing and from this dependence the nature of the state-agriculture relationship is defined. State farm pricing policy is in the service of protecting the financial interests of the state and in the extraction of capital from the countryside. Industry is protected through subsidization that occurs in the difference between the higher state purchasing price of agricultural materials and selling them to industry at a lower cost. The lower selling price was achieved because the state could absorb the transportation, storage, and handling costs for the agricultural products. In the 1980s one-fifth of the state's expenditures were for urban subsidies and were seen as necessary for maintaining

---

22. Ibid. See also Terry Sicular, "China's Agricultural Policy During the Reform Period," 340-362.

23. There are two main factions of reformers, conservatives and liberals. There is not a difference about the need for reform, but rather over the pace and nature of reform. Chen Yun, until his resignation in 1987 at the same time as Deng Xiaoping from the Central Committee, was the chief advocate for a more conservative pace of reforms. Under this view the state must remain dominant politically and economically to preserve the basic tenants of socialist development. The market place is viewed from a conservative perspective as only a supplemental form of allocation. More liberal reformers want a greater emphasis placed on the use of the market, placing the state economic role in subordination to it. More liberal reformers have been Hu Yaobang, Zhao Ziyang. Deng Xiaoping has protected and supported many of the more liberal reform ideas, though he has had to find balance between the two factions. Harry Harding, *China's Second Revolution: Reform After Mao* (Washington, D.C.: Brookings Institute, 1987), 77-83.

social stability in the face of urban sensitivity to and unrest over inflation. Hyperinflation was a major component of the 1989 student protests.[24]

The conservatives believed that the money spent on farm subsidies would be returned to the state through increased industrial output. Inexpensive food would decrease the wage bills for factories and so leave them with higher profit margins. The higher profit margins provided the state with an opportunity to take a greater percentage of those earnings through taxes and direct profit remissions. Production costs were also lowered with cheap raw materials such as tobacco, cotton, corn, hemp that were used for export or with higher consumer prices sold back to the farmers. Before 1989, to compensate for state subsidies to urban areas the policy was to cap procurement prices paid to farmers. After 1989 and the obvious willingness of the government to use military force to maintain social order at Tiananmen, the state felt that it could raise the selling price of grain in urban areas to reduce to cost of urban subsidies by the central budget.[25]

As the farmers began earning more money from higher state procurement prices in the late 1970s and early 1980s Chun Yun advocated the implementation of an idea that could be found in the theoretical writings of Mao Zedong. As peasant income increases, industry should begin to produce more consumer goods for the rural markets as a means of extracting more capital from the farmers back to the state.

> So long as the vast number of peasants have a higher consuming power, the state need not worry about having enough money to use. While consuming power is increased a large amount of money can be withdrawn from circulation in order to accumulate more capital for industrial and urban development purposes. We can practice the policy of having low prices in one sector and high prices in another sector. To sell (farm inputs) at low prices is to

---

24. Thomas Gold, "Urban Private Business and Social Change," in Deborah Davis and Ezra Vogel, eds., *Chinese Society on the Eve of Tiananmen: The Impact of Reform* (Cambridge,Mass: Harvard University Press, 1990), 176.

25. *New York Times* (March 22, 1992), 4, in Kelliher, 146.

patronize production, and to sell (consumer goods) at high prices is to get back more money for the sake of accumulation.[26]

The effect of state procurement of goods at low prices was not lost on the farmers who saw it as the tax it was. Although formal taxation actually declined, from 10% in the 1950s to 4% in the 1980's, the state procurement quotas coupled with the increased costs of consumer goods and agricultural inputs such as fertilizer kept peasant incomes low. The effects were especially felt in grain production which dropped off after the record harvest of 1984.[27] The lack of economic incentive to grow grain, the annual loss of 200,000 to 300,000 hectares (1 hectare equals 2.47 acres. So, the annual loss would be 494,000 - 741,000 acres, or up to 1,158 square miles which is an area approximately the size of Rhode Island) of farmland to urban and industrial expansion, annual population increases of 10-15 million, and increased grain consumption directly by consumers and indirectly through meat, alcohol, pharmaceuticals, and other industrial products all served to strain China's ability to produce sufficient amounts of needed grain.[28] For liberal reformers, led by Zhao Ziyang, the effect of the pricing reforms was to prove their belief that peasants are sensitive to price incentives and market signals. The liberals believe that grain should also be on the free market so that peasants will not have to be coerced and the urban population will help support the costs of modernization. The conservatives, led by Chen Yun, on the other hand, saw a lack of state control and the potential for

---

26. Chen Yun,"Chen Yun's Speech at the CCP Central Committee Work Conference," translated in *Issues and Studies* 16 (April 1980), 95, in Kelliher, 150.

27. William Hinton believes that this record harvest was the result of farmers selling stored grains given to them when the collectives disbanded, rather than the result of having produced a large harvest. This would also account for the significant decline in production the following year in 1986. Hinton, *The Great Reversal*, 22.

28. Tian Jiyun, "China's Current Agricultural Stiuation and Policy," *Beijing Review* (January 8-14, 1990), 19 in Kelliher, 163.

social upheaval. In September 1985 Chen Yun advocated the reimposition of state planning as a reaction to the liberal's procurement reforms in January 1985.

The liberal's price reforms did away with procurement quotas and installed only negotiated contracts. The contracts were negotiated between party cadre and the farmers to allow peasants a choice with regards to the quota amount or whether or not to enter into a contractual agreement at all. The return to state planning also led to the use of coercion to force the farmers to sell grain to the state. Coercion by local leaders came largely in the form of challenging access to luxury items or institutions, for example, threatening to pull children out of school, confiscating televisions or other appliances, and in some cases calling out the militia to ransack homes for grain.[29] The use of coercion also came up in my own interviews in Jiangsu province. A hamlet leader was telling me that he used his authority to determine what market a given family could sell on, the government or free market, depending upon whether or not they followed hamlet rules about crops planted and birth control.[30] Ultimately the government continues to be the only large scale market for grain and in this way maintain(s) control over the peasants to minimize peasant power over economic decisions that can affect the financial and developmental plans of the state.

Pricing and procurement policies in China cannot be divorced from political concerns about stability and growth, both of which are a function of national goals and power relationships between agriculture, industry, and the state. The achievement of goals is based upon the outcome of factional infighting among CCP leadership and factional perceptions about how best to balance the three-way relationship between the state, the farmers, and the industrial development so as to maximize growth without jeopardizing social stability. The concern for stability is particularly acute with regards to the urban populations

---

29. Kelliher, 165-66.

30. Author interview, November 1994, Dafeng, People's Republic of China.

82

who are very sensitive to changes in costs of living and are accustomed to government subsidization of food and manufactured goods.

What may appear to the Western observer as positive steps towards liberalization in the form of price reform must also be considered in the context of the CCP's views regarding the control necessary to maintain the social stability that will facilitate achievement of strategic development goals. The difference in these views is between a concern for implementing the 'right' processes and the concern for doing what is necessary at the moment to achieve specific goals. Political power is a matter of degree as to whether it lies primarily with those involved in a process or with those involved with defining and setting goals. By focusing so strongly on changes in the political and economic processes in China, American analysts want the locus of power to shift more from a centralized leadership to a more interactive market or democratic process. However, if one sees price reforms not as liberalization but as a strategic move to maintain stability, power is not shifting. Instead, new means are being adapted to achieve defined goals; power is adapting, but not shifting. The goals of reform and development are not as much an issue among China's leaders as are the means and pace of reform.[31]

Privatization

Deng Xiaoping and the conservative reformers, based on the belief that by 1978 the level of state control had grown intrusive to the point of harming production, decided that the private citizen needed more responsibility in making economic decisions. To privatize agriculture the powers of the commune system were diminished with the adaptation of the responsibility system whereby land was returned to families (though legally it continues to belong to the state), and public enterprises were leased or sold to private investors. Management decisions were reassigned to farmers and entrepreneurs, and the market for goods, labor,

---

31. Harding, 77-83.

capital, and income was expanded. As Kelliher points out, privatization tends to have a snowball effect, for as sections of the economy are privatized they can not function without the privatization of other resources that the new private 'owners' need for efficient operation.[32]

Initially land contracts to families were one-year terms. Out of concern that peasants would not plant a fall crop as their contract term came to a close, the contracts were quickly extended to three years. Families treated the land as if it were their own and not state property. The land was rented, leased, or sold and labor was hired to farm it. Land was also used as collateral for loans. Building homes became the most common way for peasants to spend their income to improve and make claims to 'their' land. Building a home was an investment in the land that symbolized permanence, wealth, inheritance, and independence, and not insignificantly the new homes improved the quality of life.

> Houses were the only important form of private property remaining to the villagers under collectivization, and they have come to serve as the major expression of economic status. The house is the son's most significant inheritance, without which he can not find a wife.[33]

More money has been spent on home building than has been given to the collective for reinvestment in land nutrients, construction, waterworks, and general maintenance. In 1980 housing expenditures reached 17.5 billion yuan and 10.5 billion yuan was given to the collective. One year later in 1981, 21.5 billion yuan were being spent on housing and only 8.9 billions were being given to the collective.[34] Again, to encourage investment in agriculture the government extended the contract period from 10-15 years.

---

32. Kelliher, 177.

33. Potter and Potter, *China's Peasant's*, 222-23.

34. Ibid., 181.

Another effect of decollectivization was the reapportionment of land into small, scattered plots of land. The small size of the plots (the average size is .6 hectares, or approximately 1.5 acres) prevents any kind of efficiency that may be achieved by growing crops on a large scale. It limits the use of machinery and increases the time and labor required if different families are growing in small quantities what one family could grow if they had a large enough area. Farmers began to rent land in order to have larger, more efficient plots to cultivate, and entrepreneurs rented out their land so that they could still have the economic security that land offers while working in business and industry earning better incomes. The result of the relative privatization of land and the increase in rural industry use were to create a need for credit and labor markets.

Credit and Labor

In 1979 the costs for farming began to be shifted from the collective to the peasant. In order to cover the costs of inputs, maintenance, and improvements in the forms of fertilizer, fuel, tractors, draft animals, irrigation, etc., the farmers needed access to credit. In 1979 the Agricultural Bank of China was activated. Over half of its loans were for the procurement of the harvest by state agents, the next largest majority went to township industries and business, and only 5% went to farmers.[35] Credit cooperatives were intended to provide credit loans to family farmers. However, they remained under the control of the Agricultural Bank. The funds in credit cooperatives were primarily from peasant's interest-bearing time deposits, but the cooperative did not control the use of the savings. The Agricultural Bank required that 30% of the cooperative funds be placed in the bank as reserve funds. The farmer's money was further removed from the rural areas in two ways: first, the Agricultural Bank was required to keep reserves in the Bank of China, and second the Bank's loans using this money were to state

---

35. On-Kit Tam, "Rural Finance in China," *China Quarterly* 113 (March 1988), 67, in Kelliher, 191.

agents, businesses, and industry. The cooperatives had to keep 5-8% of their funds on hand for cash withdrawals, which further limited their loan making ability.

With limited funds credit limits were soon reached, and in 1985 the state imposed stricter controls on rural loans to tighten the money supply.

> Banks and rural credit cooperatives only accepted deposits but did not issue loans. In some areas, credit organizations were on the verge of paralysis, and circulation of currency nearly came to a standstill. The whole rural economy sustained heavy damage.[36]

In 1989 credit ceilings were place on credit cooperatives further limiting their lending capabilities, and all excess funds had to be turned over to the Agricultural Bank. Though credit was restricted the need for money was not abated and a private credit/loan market emerged. In spite of higher interest rates charged, 4-5% a month compared to the cooperative's 0.3-0.7% a month interest rates, private markets continue to flourish accepted by the government as a necessary evil.[37]

The changes in the forms of farming also gave birth to large numbers of unemployed or, in more positive terms, a labor surplus. Some farmers with too little land to make a living farming and not enough money to rent more land contracted out as farm labor, whereas others decided to go to the cities and earn big money in business and industry. In the 1980s 100 million people fell into the category of surplus labor.[38] One option for these people is to move to the already crowded cities in hopes of getting higher paying jobs. Between 1983 and 1985

---

36. Gao Hongbin et al., "Chang xian zengzhang, yihuo fazhan chizhi" (Normal growth or sluggish development), *Jingji Yanjiu (Economic Research)* 9 (1987), 50, in Kelliher, 192.

37. Kelliher, 195.

38. Kelliher, 196. Chinese scholars estimate that one-third of the farmers are underemployed, meaning they are only needed at peak planting and harvest times. Judith Banister, "China's Population Changes and the Economy," in *China's Economic Dilemma's in the 1990's*, 242. See also Lang Deng, *Internal Migration and Development of China, 1982-1987*, (Masters thesis, University of Utah, 1991).

86

the percentage of the population classified as rural dropped from 783.69 million or 76.5% of the total population to 662.88 million or 63.4%. This is an increase of over 100 million people in urban centers in a two-year period. The shift of people from rural to urban areas is continuing to the present.[39]

Two options existed for dealing with the problem of a labor surplus, to redistribute the land into more efficient, large scale farming plots managed more as cooperatives rather than the old collective system, or to encourage small scale household entrepreneurship called specialized households. A specialized household is a home run business, initiated and run by the family. In 1983, having recently decollectivized farming and wishing to avoid that sensitive issue, the government gave strong support to specialized households. It is interesting to note that many households had already been quietly engaged in entrepreneurial activities since 1978. This is again a case of the government responding to already existing conditions in the rural areas.

Entrepreneurs became the new heroes of Chinese development. The 10,000 yuan households (approximately ten times the average annual household income) were the new models of success. In 1982, the party propaganda chief Deng Liqun stated:

> with the great changes that have occurred in the countryside, it is no longer appropriate to continue the class line of 'relying on the poor and lower-middle peasants.' What would our work line be then? Some comrades have suggested that we rally and organize the advanced element in the villages and rely on them.... I think that until we have a better method we should try this (entrepreneurship).[40]

---

39. Lang Deng, p.18.

40. Deng Liqun, "Jiaqiang he gaijin nongcun sixiang zhengzhi gongzuo de ji dian yijian" (Some Opinions on Strengthening and Improving Rural Ideological and Political Work), *Jingji Yanjiu Cankao Ziliao (Economic Reseach Regarding the Means)* 1 (January 1, 1983), 16, in Kelliher, 225.

Successful farmers and entrepreneurs were perfect models: other peasants could be expected to envy their wealth and emulate their success. ... the state could assume that these richer households were politically dependable (owing their success to government support) ... and finally, these technically skilled, better-educated people represented an alternative to the older cohort of rural cadres.[41]

By 1984 the government claimed that these 10,000 yuan households represented 14% of the population. Unfortunately, due to the heavy pressure upon the cadres to promote this level of successful entrepreneurship many success stories were staged or faked for the media and official visits by the local cadre. In 1985 the figure was revised to a more moderate 2.3% of the population.[42]

The strong state support for entrepreneurship had several effects. Socially it created more cleavages: cleavages between the haves and the have-nots; between educated, skilled workers and uneducated, unskilled laborers. There were/are owners, renters, leasors, laborers, state farm workers, rural/urban, employed/unemployed.[43] I was told by a university professor that currently in China there are officially six social class categories of peasants, and only the highest is referred to as a farmer. All of the different classes are based on levels of income, land owned, and work abilities. The political and social struggle in China now is over the difficult question of how to control and manage these new and unwieldy sets of relationships.

The commune system offered clear lines of authority that lie in sharp contrast to the now decentralized lines of power and decision making. Collectivization was a means of establishing state control in every village in China for the first time in history, and through this control it established the

---

41. Kelliher, 226.

42. Ibid., 202-203.

43. Huang, 290-300; Potter and Potter, 307-312; and Azizur Rahman Khan et al., "Household Income and Its Distribution in China," working paper in economics, Department of Economics, Univeristy of California Riverside (December 1991).

central government, through ideology, as a competitor with the peasants for control of the harvests. After the successful establishment of state power through collectivization the state now relies on and favors through supportive policies entrepreneur farmers and businessmen proxies to oversee the work of national development. It is because of the entrepreneur's dependence upon the state for political and economic support for their own success that they in turn are beholden to reciprocate support for state structures.

> The state may become particularly attentive to groups promoting commercialization and economic revival, and less sympathetic toward peasants pursuing a traditional farm life that reformers regard as a dead end. ... in the 1980's, China demonstrated that privatization is not merely a tool seized upon to repair the economic woes of state socialism, but a political choice, embraced because it meets the needs of the evolving state.[44]

There is a tension between state control and peasant independence that since the beginning of the reforms has been a relationship trying to find a comfortable balance. Peasants can exert power in several ways: (1) they can manipulate policy either through a deliberate misconstruing of the policy or by exploiting weaknesses in the new rules; (2) they can break the rules as a means of experimenting and testing new policies; (3) they can achieve necessary production by using policy alternative banned by the state; and (4) they are aware of the effectiveness of passive means such as sabotage or deliberate lethargy.[45] On the other hand, the state does at present maintain ultimate power over the population at large. The effectiveness of peasant exertions of power is not an overall change in the policy itself, but can be a qualitative change in the degree to which the policy is effective or encompassing, for example, the decision to implement the household responsibility system, or the initial manipulations of the pricing system. It is the important role of agriculture as a source of raw materials, money,

---

44. Kelliher, 232.

45. Kelliher, 239-242.

and inexpensive food necessary for industrial and urban development that gives peasants the leverage that state leaders must and do listen to. The interdependent relationship between industry, agriculture, and the state, combined with the twin imperatives of achieving development goals while maintaining social stability, creates a situation that requires careful political balancing. Sudden changes or shifts of power in any one area have ramifications for all of the others.

In 1986, when ecological agriculture first officially came on the scene, social divisions and alienation from the process of development were an increasing problem. With the increasing social divisions between the urban and rural areas and increased populations shifts to the cities and the 'dead-end' work of traditional farming, there was a need to stimulate motivation for farming as profitable way to earn a living. These are issues of social stability as well as of production. With increasing dissatisfaction there is an increased potential for both open or passive acts of protest that will have a negative impact on the need for constant, efficient production. One of the political effects of ecological agriculture is to address government concerns towards dissatisfaction by trying to encourage farmer cooperation. As seen in the introduction to the policy below and in greater detail in subsequent chapters, the language used to describe the method and philosophy of ecological farming is built upon the ideas of integration, harmony, and using existing resources to improve production and quality of life for participating farmers.

It is also interesting to note that ecological agriculture began as an environmental project. The pressure to increase production in the face of a rising population and shrinking land resources is very real. Vast areas of land have been deforested, lakes have been dried up, any and all 'unused' land is being turned to farmland. On top of this, farmers are attempting to produce more on the same area of land with the increased use of chemical fertilizers. Land reclamation and

90

chemical additives have had limited success, but it has also been more destructive of resources that might have been better used or more wisely managed.[46]

The environmental situation was explained to me in the following way.

> In the 1970's brigades did not use too many chemicals. They relied more on green manures and pond bottom slurry, and at that time soil fertility was not too bad. My parents had a good understanding of organic agriculture because they did this before the 1960's. They stopped because rural enterprises were producing much higher profit products. It's said that one day in enterprise is equal to three or more in agriculture. Organic (ecological) agriculture is labor intensive, so if people use it it can absorb the surplus labor. But, people are lazy and they want the agricultural production without too much work so they buy chemicals. They are easy and handy to use, and now it is a big problem for my home town.
> My father visited me and told me that in my home town (in eastern Jiangsu province) agriculture has a big problem, soil fertility has decreased. In the past the good soil was up to 18 inches deep, and now it is only 4 inches. Also the animals such as snakes, frogs, eels, crabs are all decreasing. When I was a kid (1970's) I could catch many in the rice fields and now I catch none. According to my father this is due to the heavy input of chemical fertilizers and pesticides. In the past there were only a few cases of cancer and now there are many cases of cancer. My father came to Nanjing for stomach cancer - I tried to convince him not to drink from the river, so I convinced him to dig a well. Unfortunately, it is expensive to dig a well and he only dug 5 meters and the water is still not so good. The good water is 50 to 100 meters deep.[47]

---

46. Cheng Xu, et al., "Sustainable Agricultural Development in China," 1128; Vaclav Smil, "Land Degradation in China: An Ancient Problem Getting Worse," in Blaike and Brookfield, 214-222; and He Bochuan, 21-42 and 98-104.

47. Author interview, November 1994, Nanjing, People's Republic of China. In separate interviews, one with an American conducting a study of nitrogen flow pattern, and another with a Chinese agronomist, I was told how fertilizer application is haphazard at best. Often it is tossed onto the crops by hand rather than in measured even applications, and it is applied when it is 'believed' to be necessary. The result is poor coverage of the area and clumps of fertilizer in small spots. These small clumps quickly saturate the soil without dissipating into a wider area, and up to 85% of the amount applied passes through the soil unabsorbed, into the water supply.

Ecological agriculture as a policy works with existing peasant knowledge of farming practices, which is a mix of traditional and modern methods. It brings past farming experiences into the present. it emphasizes that hard work in agriculture can pay off, and it offers a way to correct what the people can see happening to their villages in terms of environmental degradation. Ecological agriculture is based on familiar methods of nonchemical, labor intensive farming, and in its model-based form of implementation it adheres to the two fundamental principles of rural organization; "peasant participation must be voluntary, based on the economic success of local models, and the principle that income must be distributed on the basis of work performed."[48]

The Technique and Philosophy of
Ecological Agriculture

Recent Chinese experiments with ecological agriculture (CEA) address the problems of production and environment while working within the existing economic, land, and labor policy constraints. CEA uses crop rotation patterns, stereo-cropping, animal and green manures (traditional agriculture methods), supplemented by chemical fertilizer and chemical pesticides (western technology) for maximum use of land, optimal use of seed and crop types to take advantage of seasonal changes and minimize disease and pests. Ecological methods do not eschew chemical additives but, instead, emphasizes their use only as a supplement rather than as a primary fertilization and pest control as has been encourage for the last thirty years. Additionally, cash crops are intercropped with food crops to allow farmers greater opportunity to increase their income.

All human, plant, and animal wastes on the farm are recycled if possible. Animal wastes are put in fish ponds for fish food; sludge from the ponds is used for field fertilizers. On some farms all animal and human wastes go into a bio-gas

---

48. Hinton, 51. These principles are not only visible in the implementation of ecological agriculture, but they can also be seen in the selection of new farmers and entrepreneurs as models, who will ideally stimulate voluntary emulation and income based on work done.

92

digester. With one cow, two pigs, two chickens, and five people enough methane can be produced to provide the household with gas for an entire year. The methane is used for cooking, heating, and electricity and has the added benefit of improving sanitary conditions in the villages. The slurry from the digester is applied directly to the fields as fertilizer and has even been experimented with as pig food. The pigs loved it and even became intoxicated by it.[49]

The design of the ecological model is based upon the symbiotic nature of all living organisms, including people.

> In the eco-system, living organisms do not exist in separation, but are closely connected with the environment surrounding them. They interact with each other and coexist within a unified whole. There is a complicated relationship of material and energy exchanging between organisms and their environment. On the one hand, in order to exist and multiply, organisms often have to absorb material and energy from the environment such as air, water, light, heat and nutrients. On the other hand, during the process in which organisms exist, multiply and conduct activities, they constantly release and excrete materials, and give their remains back to the environment. They are thus in a complementary relationship with the environment. The Environment influences organism and organisms, in return influence the environment. The changed environment under the influence of organisms will, on the contrary, influence organisms, hence there is a process in which they constantly interact with each other and evolve in a co-ordinated way. As far as this relationship between them is concerned, organisms are occupants of the environment, (and) they are the component part of the environment in which they exist at one and the same time. As occupants, organisms constantly use environmental resources and transform the environment, while as a component part of the environment, they constantly compensate the environmental resources so that the environment will maintain a certain amount of materials on reserve to guarantee the re-birth of organisms. Adhering to this principle, eco-farming stands for rational arrangement of production, rational rotation of different crops in the light of actual conditions and at the right time, and the method of growing corps on some plots of farmland during one period and keeping them idle during another.

---

49. Jiangsu Province Department of Agriculture extension worker interview with author, November 1994, Nanjing, People's Republic of China.

> Should eco-farming run counter to this principle, it would lead to
> the deterioration of the quality of environment and the exhaustion
> of resources.[50]

There are three key concepts based upon the multidimensional use of space and time that make up the planning of CEA and from which more sophisticated models of integration are formed. One can be called horizontal, which is fixed linearly on the earth and includes methods of intercropping, relay cropping, and crop rotation based upon specific planting and harvest times. A second and complementary set of considerations are vertical. This is referred to in the ecological literature as stereo, or multistoried, agriculture and involves making use of up and down spaces given different species requirements. Third, is the practice of recycling, which will be discussed below. These practices are not new to Chinese agriculture, but the discussion of them and their effects as a holistic system with environmental and economic benefits gives them a value and importance relevant to the process of modern development. I will briefly define four of the key terms for both horizontal and vertical perspectives (intercropping, relay cropping, crop rotation, stereo agriculture) and then describe several different situations to illustrate their implementation.

The term *intercropping* refers to the practice of planting more than one crop in a field at the same time. For example one can grow three different crops in a field at the same time at different stages of maturity. For example in between rows of cotton may be planted rapes (an herb of the mustard family grown as a forage crop for sheep and hogs, and its seeds yield rape oil and are bird food), and vegetables. Another combination I was told about is the planting of barley, rice, and corn, or soybeans. Relay cropping involves planting a new crop ten days to two weeks before the harvest of an existing crop and can be done in conjunction

---

50. *China's Eco-Farming*, compiled by China's National Environmental Protection Agency, and the United Nations Environmental Programme, (Beijing: China Environmental Science Press, 1990), 12. In addition to the book, *China's Eco-Farming*, information in this section comes professional articles made available during my stay in China, from personal interviews, and from Cheng Xu, et al., "Sustainable Agricultural Development in China," 1127-1144.

with intercropping. In this way cycles of growth and harvest can be taken into consideration to maximize the farmers use of land and time. Crop rotation is the practice of planting different crops in alternate years to allow the soil an opportunity to rest and regenerate. Cotton is particularly demanding on soil nutrients, and planting a field with legumes in an off year can return lost nutrients to the soil and give the soil a break from intensive fertilization and irrigation required by cotton growth. In stereo agriculture more consideration is given to the height and sunshine requirements of plants and animals. One may plant tall trees such as Asian pears or rubber trees with medium- sized trees such as peaches or shrubs such as tea. In appropriate areas this can also include planting edible fungi. Stereo cropping can also occur in rice paddies. Rice can be grown with azolla (a nitrogen-fixing aquatic plant historically used in China as a green manure) and with fish in the paddy as well. Some other combinations include planting rice with algae and fish or planting rice with ducks and fish, or trees with ducks and fish. Even within ponds different species of carp are raised that feed at different levels of the pond. Below are two examples of how intercropping and stereo cropping have been successfully applied.

> The Beichuan County in Sichuan Province tried to inter-crop forest trees with crops from 1981 to 1987. The county experimented on 9,866 hectares of land, planting 860 hectares of tung trees inter-cropped with corn and soyabean, 2,924 hectares of fruit trees with tea and mulberry trees or grain crops, 1,030 hectares of forest trees with rapes or sunflowers, taro and other cash crops, and 150 hectares of forest trees with medicinal herbs. In these years, the county increased grain output by 1.15 million kg, oil-bearing crops by 37,000 kg, medicinal herbs by 36,000 kg, vegetables and taro by 559,000 kg and generated 1.12 million yuan in income.

> The Daqiao Township in Jiangjin County, Sichuan Province, raised 1,783 duck on 6.7 hectares of paddy-rice fields in 1988. The township raised duck in the rice fields for 84 days from May 10 to July 31 and harvested 3,031 kg of adult ducks, generating 10,900 yuan. Also, the township harvested 3,150 kg of fish and took in

12,600 yuan from the fish in the rice fields. At the same time, the grain output increased by 3,050 kg that year.[51]

The third component of CEA, the one that binds together the horizontal and vertical methods of cultivation and integrates animal and plant growth, is the recycling and utilization of waste materials. In a simple system scraps of bark and wood chips from trees could be used to produce edible fungus, and the discarded material from the fungus beds in turn would be used as fertilizer for the trees.

In more complex systems of integration, plant and animal manures are used for fertilizer, food, and methane gas production through the use of bio-gas digesters. Pond and stream sludge are used to fertilize fields, chicken droppings can be used to feed pigs, pig and duck droppings feed fish, and the fish droppings and uneaten mammal droppings generate fertile pond sludge for the fields. With a bio-gas digester animal and human droppings can also be used to generate methane gas for people, and the dregs of this can be used for fertilizer or, as mentioned earlier, pig food. Again, examples are more illustrative of the process of integration and recycling.

> Formerly, Wang Shouqing, a farmer in Dafeng County, Jiangsu Province, was primarily engaged in crop raising and animal farming. Grain produced from his farmland was directly used to feed chickens and pigs. The result was that his cost was high and low profits. In 1982, he set up a starch workshop and raised silkworms. Broad beans were processed into vermicelli (made from bean starch) which was sold on the market. Leftover material is quality feed for chickens and pigs. Also, silkworm excrement and chicken droppings are [*sic*] also used to feed pigs. As a result, the feed cost for pigs was reduced by 50%. In addition, Wang bought two donkeys for the starch workshop. Donkey droppings can be used to feed fish. Part of the donkey and pig droppings can be used to feed fish. Part of the donkey and pig droppings was used on the farmland, thus reducing chemical fertilizer. In 1983, Wang netted more than 16,000 yuan in profit.[52]

---

51. *China's Eco-Farming*, 36 and 38.

52. Ibid., 43.

> The Guquan Ecological Farm in Nanjing has rationally combined the production of pig and fish farming, the biogas project, the raising of ducks, the raising of earthworms, the growing of mushrooms and fruit trees into an integral sound recycling system. Chicken droppings are used as part of the feed for pigs. Pig droppings in the biogas pits are used as raw materials for fermenting biogas and the generated biogas used for cooking and lighting. Biogas liquid and dregs can be used to feed fish, cultivate edible fungus, raise earthworms and as fertilizer for orchards. The fish pond also provides a space for raising ducks. Duck droppings can be used as fish bait in the pond and the pond silt can be used as fertilizer for the orchard. Earthworms are raised in the orchard, droppings and mushroom dregs can be used as bait for earthworms. Earthworms can improve the orchard soil (aeration) and earthworm droppings provide fertilizers for the orchard. Earthworms can be used as the feed for chickens. The various production systems of the farm are co-ordinated with each other so that their productivity is comparatively high and able to yield sustained growth.[53]

The integration of animals and plants has also reduced the need for pesticides as well as fertilizers. Chickens raised in cotton fields control bollworm and scarabs, and ducks have the same effect in rice paddies against a wide variety of pests such as rice leaf rollers, locusts, and paddy borers. Dense plant coverage with beneficial plants and ground covers also controls weeds, preserves the soil and enhances soil fertility. The integrated nature of CEA and the attention to detail required in the making of ponds, fields, and orchards also control soil erosion, and desertification (the product of overused and improperly used land) and improves overall soil quality

Three-dimensional cultivation, mixed animal farming, plants mixed with animals, recycling of resources, weed and pest control, environmental benefits (desertification control, soil erosion, soil quality) are the success stories of CEA. These are methods that have a recorded history in China of hundreds and in some

---

53. Ibid., 50.

cases thousands of years. During the Jin (265-420) and Tang (618-906) dynasties it is written that red tree ants were used to control citrus pests in Guanzhou. What are now considered to be the traditional methods of farming took shape in the Han (206 BC -220 AD) and the Northern and Southern dynasties (420-589).[54] These methods continued to be widely used in China until the mid-1960s and are still largely familiar to farmers.[55]

In the late 1970s Chinese researchers, borrowing the idea of sustainable agriculture from the United States, supported the growing Western reaction against the chemical-based farming and its negative effects. A series of conferences, including one in 1982, were jointly sponsored by the Chinese Office of Environmental Protection and the East-West Center of the University of Hawaii. Pilot projects were started in the early 1980's and official backing for national implementation came in 1986. In reacting to Western chemical or science-based agriculture, a vision of development was articulated that played on many political leaders' concerns about becoming overly identified with Western practices or of becoming Western at the expense of Chinese culture and history.

The context of Chinese ecological agriculture is one that is both historically Chinese and a product of the reforms that occurred after 1978. As was illustrated in the examples above farmers were able to take advantage of the ability to contract land and increase personal profit, activities not previously possible under Mao. The philosophy of ecological farming recognizes that the individual farmer is never completely alone in all production decisions. There remains a broadly based scale planning role for governments at all levels to coordinate the use and administration of resources. This can be seen in the county and township levels of implementation of CEA, and the importance of a government role is explicitly mentioned in the literature that will be discussed in the next two chapters. There is an explicit and implicit recognition of the larger

---

54. Ibid, 55 and 4-5.

55. Based upon author interviews Nanjing, People's Republic of China, November, 1994.

social and economic perspective that progressively higher administrative levels have of resource needs and production abilities and of the important relationship between the individual and the whole.

Lastly, the ecological farms and farmers in the examples above and all pre-1986 examples are designated as models. As official models they are supported and financed through university and government offices to experiment with the processes and principles of ecological agriculture. We can see from the examples above that there is a requirement for monetary investment to improve ponds, paddies, and fields and to purchase more animals, fishes, and seeds necessary to diversify into a more labor intensive type of farming. The payoffs appear to be high, but as will be explored in the next chapter they are financially out of reach by the classes of peasants that need them most, those that are not able to leave the land or able to earn money from farming. With entrepreneurs and large successful farms as the new model of success in the 1980s and 1990s, larger percentages of people moving to the cities, and larger numbers of people considered as underemployed or simply surplus labor, there needs to be a model of farming that makes agriculture an attractive and realistic possibility.

The government has a stake in keeping agricultural production in pace with industrial development, for achieving national goals and maintaining social stability in both urban and rural areas. The first section of this chapter attempted to show the interactive nature of the peasant/government relationship in China. Of specific interest for this book are the exercises of power. On the part of the government power is expressed through the formation of goals and rules/policy for achieving those goals. The peasants' power lies in their interpretation of the rules and their resulting action. The rules can be followed, openly disobeyed, or passively dealt with as a means of protesting and as a means of testing to see what can be gotten away with. Some rules, such as instituting family farming, are vague and left to local interpretation as a means of implementation. With other rules however, such as pricing policy, government leaders are less tolerant of creative interpretation and change the rules to counteract peasant interpetations.

What is not changing in this interaction are the goals. Instead, there is a play of rule making, interpretation/action, then followed by another rule responding to the action. There is an ongoing tension between the government and the farmers in the process of moving towards the government goal of increasing national agricultural production and income often at the expense of improved farmer incomes and standards of living. Ecological agriculture as a philosophy describes a way of working and interacting, a set of roles and rules that are relatively more integrated and harmonious, a way that in its ideal expression has the potential to ease existing tensions.

One goal of this book is to understand ecological agriculture not only as a successful method of agriculture in China but to see it in the context of political and cultural meanings that are engendered by CEA as a symbol of China's current development dilemmas. These dilemmas are (1) the economic tensions between a need for ever increasing production in spite of limited, diminishing resources and (2) a cultural identity crisis reflected in the increasing social divisions that are the result of rapid economic growth and the exchange of ideas and technologies with the West. The language used to discuss and implement the policy of ecological agriculture, as analyzed in the next chapters, gives one perspective of a means of dealing effectively with these dilemmas. It is a perspective that has a culturally strong enough resonance to have facilitated its nationwide implementation.

CHAPTER 4

ECOLOGICAL AGRICULTURE AS A POLICY

This chapter will examine the symbolic components of the Ecological Agriculture policy as follows: (1) the program name, "ecological agriculture" as a metaphor that shapes discussion, reflects cultural thinking about agriculture and directs action, (2) the use of goals and models in policy implementation as a culturally familiar ritual action that actualizes the myths which are categorized and supported by the problem-setting function of the metaphor, (3) the discussion of ecological agriculture as myth that overlooks policy contradictions and generates quiescence through the language used to discuss and resolve the problem. The chapter concludes with a summary of the major themes as embraced by the concept of model-following. This is the primary concept to emerge from my research and due to the lack of appropriate terminology in the literature or from my interviews I have labeled it myself as model-following.

What's in a Name? Ecological Agriculture as a Metaphor

Metaphors allow one to "comprehend one aspect of a concept in terms of another."[1] They serve a range of actions such as referring, quantifying, identifying, and setting goals and motivating action. The way one interprets the world is based upon the metaphors one uses to define and describe what one sees and hears. To see the program title as a metaphor is to ask about what possible

---

1. George Lakoff and Mark Johnson, *Metaphors We Live By* (Chicago: University of Chicago Press, 1980), 10.

meanings it has for referring, for setting goals, and for motivating action. As a metaphor ecological agriculture combines two uncommonly related concepts, that of ecology and agriculture, to create a third image or notion about agriculture. Agriculture is modified by the idea of ecology, and it can be stated as an equivalency that agriculture is ecological. The notion of what is ecological is further elaborated by equating it or making it inter-changeable with concept of Chinese characteristics. The following analysis will examine the concepts of agriculture and of Chinese characteristics combined in the metaphor of ecological agriculture and show how they reference fundamental social values of social and environmental integration and extend their relevance to new integrated methods of agricultural modernization. The result is a culturally specific metaphor that is grounded in the Chinese historical experience and shapes thinking for policy actions.

Within China what is referred to in the United States as sustainable agriculture is more commonly referred to as ecological agriculture (sheng tai nong ye), which was described to me by a leading agronomist and one of the founding researchers of ecological agriculture as not only a method but a philosophy of agricultural development. In his words, it is, "agriculture with Chinese characteristics." Ecological agriculture is a metaphor that combines notions of what is Chinese through the term *ecological*, with the concept of agriculture. What is 'Chinese' as discussed in relationship to agriculture? The Chinese characteristics described to me in formal and informal interviews, as well as in written texts, are brief descriptions of what are seen as the significant historical, social, and physical conditions that have shaped their agricultural experience.

The first item always mentioned is China's burgeoning population. Population density has been an agricultural consideration in Southern China since the Southern Song Dynasty in the twelfth century AD, at which time integrated agricultural methods began to be practiced.[2] The National Population Growth Planning Commission estimates that the population will reach 1.6 billion before a zero growth rate is achieved. The current growth rate is approximately 17 million people annually, and the Chinese Science Academy has stated that China's arable land will sustain at most 1.5 - 1.6 billion people.[3] One may ask how this is a characteristic, and the answer must be that it is a Chinese characteristic not because it is unique to China but accept it as a part of often repeated self-definition that has shaped Chinese identity. It is a historical issue that has specific relevance for the role of the farmers as food providers for a nation.

The second characteristic is a limited resource base that must be protected from the intense use of so large a population to support the same and growing population. China is the world's third largest geographic area, with approximately 21% of the world's population to be sustained on only 7% of the world's arable land, and one-fourth of the world's fresh water. Within China the richest resources are in the eastern third of the country which is also home for the majority of the population. The amount of land and water available for agricultural use is diminishing rapidly due to the increased use of land for living areas and for commercial development. In recent years Beijing has had to ration water, and wells in the surrounding area are going dry due to the demands of the city. Water and soil pollution have also had a negative impact on the agricultural and urban areas. In Shanghai the cleanliness of Suzhou Creek is measured by the number of days a year that it is not black, now about thirty days a year.

The reiteration and emphasis placed upon historical roots focus attention on the unchanging character of China's social and environmental dilemmas, and

---

2. Wang Zhaoqian, "Historical Cultural and Social Backgrounds of Chinese Ecological Agriculture," paper presented at the World Sustainable Agriculture Association conference, Beijing, People's Republic of China, September 1994.

3. Zhang Xigu, "The Key to the Breakthrough for Chinese Sustainable Agriculture: Organic Combination of Chinese Traditional and Modern Agricultural Techniques," paper presented to the World Sustainable Agriculture Association conference, Beijing, People's Republic of China, September 1994.

104

the ability of the people to work within these constraints to overcome hardship. The need to work within natural constraints and to renew natural resources is traced back to the philosophies of Zhuang zi and Xun zi, over 2000, years ago which gives modern ecological agriculture a strong historical foundation and precedent. Traditional approaches to agricultural production have been practiced from ancient times until the Cultural Revolution and have successfully maximized production and renewed limited resources through integrated farming practices.[4] By talking about present problems with examples from the past, a continuity is created that bolsters the sense of being Chinese, of being connected to the past through the struggle with common problems. The historical division that is culturally made by singling out the Cultural Revolution is significant because it was during the mid-1960s when chemical fertilizers became popularized, and this can be seen as a move towards a more interactive relationship with the West.

The third characteristic is the low social and economic status of agriculture. The low economic return for farmers is one of the main problems to be overcome by ecological agriculture. "The (economic) benefit of farming is very low compared with other business. China is now under a state of high production, high input and low return resulting in farmer's low enthusiasm in farming."[5] Old people and children too young to work in the factories are more commonly seen working the farms as increasing numbers of people move to the cities to find work in businesses and factories. The official and popular emphasis on business and industry as moneymakers in recent years has also led to the perceived inferiority of agriculture as a profession. Through informal talks with students at Nanjing Agricultural University, I learned that they believed they were getting a good education but mentioned that the only problem was that the word 'agriculture' would be on the graduation diploma. They felt that this would affect

---

4. Ibid., 1.

5. Ibid.

the way others viewed their education and impact the types of jobs they might find.

What is agriculture as discussed in relationship to 'Chinese'? In discussions of agricultural practices there are two general categories, Chinese traditional agriculture and Western petro-chemical-based practices.

> The hallmark characteristic of traditional Chinese agriculture as an intensive use of food-producing natural resources, with twin objectives of achieving high short-term food production levels and nurturing the natural resources so as to not impair the country's long-term production capacity.[6]

As discussed in the previous chapter, this involves the use of plants and manures for fertilization, intercropping, relay cropping, and careful maintenance of the fields to sustain maximum fertilization and longevity of use with limited resources. These methods have evolved over 2000 years of history to form "sustainable agriculture with Chinese characteristics of meticulous tillage, agriculture-livestock integration, and rotation and multicropping based farming system."[7]

Western practices by contrast rely on chemical pesticides, chemical fertilizers, heavy irrigation, and mechanization to support the high yield seeds that are used. Since the mid-1960s the use of chemical fertilizers and pesticides in China has become increasingly widespread. These practices coupled with economic and social incentives have met with strong positive results in increasing grain yields. Petro-agriculture is the term used describe chemical-based agriculture and is often juxtaposed with ecological or eco-agriculture. Petro-agriculture is based upon having special varieties of seeds that require special

---

6. F.H. King, *Farmers for Forty Centuries: Permanent Agriculture in China, Korea, and Japan,* (Madison, WI: Mrs. F.H. King, 1911) quoted in Cheng Xu, Han Chunru and Donald C. Taylor, "Sustainable Agricultural Development in China," *World Development Journal* (August 1992), 113.

7. Wang, 2.

chemical fertilizers and pesticides; it emphasizes input over holistic integration as eco-agriculture does. Petro-agriculture is considered to be 'modern' as opposed to 'traditional' because it is less labor intensive; it produces good results more quickly and it is the result of Western science.

Using methods that are the product of intense scientific investigation in an advanced country and that appear to allow people to get more out of existing natural resources is scientific. Moreover, the use of these modern scientific methods is a reflection of China's own modernity. To be a 'modern, advanced' country many feel that China must adopt the methods of already modern, advanced countries like the United States. Ecological agriculture can stake a claim to be scientific, though based largely on traditional methods, because its researchers can show scientifically why the methods work as they do and what makes them superior to petro-agricultural methods. They have evidence from experiments at all levels of farming (the soil, water, animals, maneures, and model farms, etc.) in China and in the United States that support the validity of not relying solely on chemical inputs. Using science is modern, and understanding tradition scientifically makes tradition acceptable as modern. Becoming modern and being seen by others as modern are important aspects of policy formation. One symbol of modernity in China is the automobile. The increasing number of personal automobiles is seen as a symbol of China's growing modernity in spite of the increased pollution and congestion that exist from current traffic levels. Developing the world's most modern mass transit system is not a goal or symbol of modernization.

The international concern about the relationship between people and the environment adds to the Chinese leadership's acceptability of the term ecological as modern. Environmental awareness is a recent phenomenon brought on by the severity of resource degradation in China, which is a very real threat to overall development goals, and by pressure from the international community where China's lack of environmental regulations is seen as an unfair economic advantage. The term ecological has come to be something of a buzz word, or metaphor in China for environmental issues.

I was told by the leader of a state farm, who was telling me about the "greening" of China, that Tianjin was to become an ecological city, and that each province was establishing model cities and counties that are to be "ecological." At this point I had come to know of ecological as meaning a fully integrated system between man and nature, and between farms and rural agricultural-based industries that relied on the recycling of waste products as food and fuel. I was shocked to think that this would be attempted on an entire city, what a phenomenal undertaking, how would it be done? I expressed my amazement and asked several times to be sure of what I had heard.

When I returned to Nanjing and asked several professors what exactly was involved with making an entire city 'ecological,' I was told that it is an attempt to reinvigorate enforcement of long-standing pollution control laws and to 'green' up the streets with flowers and trees. It was not the total integration that I had learned ecological meant; those that I asked were as surprised to hear this as I was. One professor told me with a rather sarcastic laugh that "If people could they would call this the People's Ecological Republic of China." For me this emphasized the point that being ecologically minded had taken on national significance but was a way of talking about pollution control rather than any actual changes that would lead to the integration of people and resources.

Traditional Chinese and more recent Marxist attitudes towards the environment have been, as in the West, based on people's need to control and dominate nature. The Chinese historian Sima Qian (202 BC - AD 9) wrote "... without people to open up the mountains and marshes, there will be a shortage of wealth," and later Lu Yu (AD 1129 - 1209) wrote "If there's mountains, we'll cover it with wheat/If there's water to be found, we'll use it all to plant rice."[8] The

---

8. Vaclav Smil, "Land Degradation in China: An Ancient Problem Getting Worse," in Piers Blaikie and Harold Brookfield, eds., *Land Degradation and Society* (London: Routledge Press, 1987), 215.

process of land reclamation peaked in the "early 1970's when at least 100 million peasants were engaged in off-season terracing of slopes, leveling fields, filling lakes and marshes and the opening up of grasslands."[9] In recent years there has been an effort to undo some of the damage done during this period with programs of reforestation and rational (matching appropriate crops to soil and climate conditions) land use.

The following quote from the preamble of *China's Agenda 21*, adopted after the 1992 United Nations Conference on Environment and Development in Rio de Janeiro, vocalizes the recognition of a need for a new approach to economic development.

> Traditional ideas of considering economic growth solely in quantitative terms and the traditional development mode of 'polluting first and treating later' are no longer appropriate when considering present and future requirements for development. It is now necessary to find a path for development, wherein considerations of population, economy, society, natural resources, and the environment are coordinated as a whole, so that a path for non-threatening sustainable development can be found which will meet current needs without compromising the ability of future generations to meet their needs.[10]

This concern for a new path towards sustainable development is an acknowledgement by China's leaders that their limited resources and patterns of use are diminishing the capacity of China's natural resources to meet the increased demands being placed upon them. This is a shift from an attitude of simply plant and use all available resources, to a perspective that recognizes that China's geographical and social diversity demands a more sophisticated understanding, one that acknowledges diversity of needs and abilities and tries to integrate

---

9. Ibid.

10. *China's Agenda 21: White Paper on China's Population, Environment, and Development in the 21st Century*, Adopted at the 16th Executive Meeting of the State Council of the People's Republic of China on 25 March 1994, (Beijing: China Environmental Science Press, 1994), paragraph 1.1.

resources to minimize waste, to achieve the desired balance of growth and resource sustainability. This lies in contrast to policies that have dictated what will be planted where, how much, and without consideration of farmer's abilities, for example, the decision that wheat should be grown everywhere in China. Forests were felled, and lakes filled up to create areas to plant more wheat. The result was not a significant improvement in grain production and a very significant loss of other resources. Although China continues to experiment with developing relatively unfertile areas into farmable land, such as red soils areas in the north and sea beach areas in the east, there is an understanding that only certain crops can be grown in these areas and that they will require special cultivation measures.[11]

The coordination of "population, economy, society, natural resources, and the environment"[12] implies an important role for a centralized government in development. It is a role that can be observed in the implementation of ecological agriculture, and it is one that is not only socially acceptable, but is desirable. The comment that "if only the government would establish good policies" was heard more than once. In an interview with a hamlet leader whose hamlet was not involved with the ecological agriculture project, his frustration with government policy became clear. The members of the hamlet wanted to diversify from only cotton production to also raising sheep, but they lacked the money to buy the necessary equipment and sheep. "If only the government would create a good policy our hamlet could be more prosperous." A good policy would be one that would allow increased use of markets for current cotton crops (one of the few crops still solely purchased by the government), increased money from the government for investment in diversification such as sheep and was part of a

---

11. Funing Zhong, Rong Wang and Shirong Jiang, "Preliminary Analysis of Seabeach Land Development in Northern Jiangsu Province," (unpublished manuscript, 1994); also T.C. Tso, ed., *Agricultural Reform and Development in China*. (Beltsville, MD: Ideals, Inc., 1990).

12. Ibid.

110

larger government plan for macro-level market and price controls that was necessary for "keeping all of the people fed."

Agriculture in relationship to 'Chinese' is also about social relationships, about power. As a goal of modernization, agricultural development is a political relationship that involves approximately 75-80% of the population, and as such it continues to be the economic backbone of the country. Additionally, China's efforts to minimize food imports and to maintain self-reliance on grain also place pressure on the farms to meet the ever increasing demands. It is a state-society relationship that has been historically and is currently referred to with the simile 'like a family' by both the Chinese and by Westerners living in China. A currently popular song in China called "Big China" has the lyrics, "We all have a family whose name is China ... our big China, what a big family."[13] To be 'like a family' implies an expectation of behavior and responsibilities on the part of the people and the government. As described to me, in many large Chinese families though there are internal squabblings and bickerings, there is also a general unity expressed through pride and defensiveness against outside interference and judgement. There are generally respect and deference to authority vested in the father and family elders, and there are acceptable and unacceptable means of showing dissent or support.[14] Perhaps our own discussions about the loss of family values are instructive here, where we lament the breakdown in the respect for authority and the property and rights of others as well as a failure to take

---

13. Steven Mufson, "Why China Talks Tough and Flexes Its Military Muscle," *San Francisco Chronicle*, 25 March 1996, sec.A, p. 8. As a further comment on the importance of the family concept in China the following quotes are helpful.
"The family shows the laws operative within the household that, transferred to outside life, keep the state and the world in order," and "The family is society in embryo; it is the native soil on which performance of moral duty is made easy through natural affection, so that within a small circle a basis of moral practice is created, and this is later widened to include human relationships in general." Richard Wilhelm, trans., *The I Ching or Book of Changes* (Princeton: Princeton University Press, 1985), 143-144. See also Lucian Pye, *The Spirit of Chinese Politics*, (Cambridge: Harvard University Press, 1992).

14. Dr. John McCoy, Joint Venture Consultant, author interview, 23 November 1994, Shanghai, People's Republic of China. Also see Lucian Pye, *The Spirit of Chinese Politics* on this point.

personal responsibility for actions. Our efforts to redefine families or to reinvigorate traditional ideals reflects divisions in understandings about the identification of, the relationship to, and the role of authority. In the American debate the simile of 'like a family' seems also to apply, defining a different type of family. Where the government is often pitted as the enemy or that its role should be in limited forms that compensate social groups for weaknesses and inequalities in the market, and access to power holders is uncertain, unavailable, or undesirable (reflected in decreasing voter turn out and the increased power of lobby groups), we show our uncertainty about the identification of, the role of, and our relationship with authority. In families of this sort in which there are increasingly unclear guidelines for interacting with power holders, people will establish their own definitions of authority and of desirable relationships with authority figures.

The analytic concepts of morality and paternalistic power are used to refer to the expected behavior in social situations that involve different levels of authority, i.e., with parents, with superiors at work, or with government officials. Further, it is the enactment of expected behavior for both the power-holder and the subordinate which is of greater importance than the level of sincerity behind the expressed behavior, providing a forum for cooperation and opposition.[15] Both of these ideas are founded on the more traditional Confucianist notions of filial piety and fraternal duty, but for my purposes here I will only address the main analytical points of morality and paternalism.

---

15. I rely primarily on Fei Xiaotong's comparative analysis of Chinese society, written in 1947, and have found his conclusions to be supported in contemporary literature some of which are listed below. Fei Xiaotong, *From The Soil: The Foundations of Chinese Society*, trans. Gary G. Hamilton and Wang Zheng (Berkeley: University of California Press, 1992). For current explications on the nature of social relation-ships in China, see Thomas B. Gold, "After Comradeship: Personal Relations in China Since the Cultural Revolution," *The China Quarterly* 12, no.104, 657-75; Kwang-kuo Hwang, "Face and Favor: The Chinese Power Game," *American Journal of Sociology* 92, no.4 (January 1987), 944-74; Zhang Shuqiang, "Marxism, Confucianism, and Cultural Nationalism," and Zhiling Lin, "Traditional, Modernist, and 'Party' Culture in Contemporary Chinese Society," in Zhiling Lin and Thomas W. Robinson, eds., *The Chinese and Their Future* (Washington, D.C.: The AEI Press, 1994); Lucian Pye, *Asian Power and Politics* (Cambridge: Belknap Press, 1985).

The Chinese basis for association is rooted in "self-centered networks of social relationships."[16] This is different from the institutional and legal basis for social relations in the West, and it is different from the misperception that the Chinese are more group oriented or less individualistic than we in the West. "Chinese society is centered on the individual and is built from networks created from relational ties linking the self with discrete categories of other individuals."[17]

> From the Son of Heaven down to ordinary people, all must consider the cultivation of the person as the root of everything.' This idea is the starting point in the system of morality inherent in Chinese social structure. Extending out from the self are the social spheres formed by one's personal relationships. Each sphere is sustained by a specific type of social ethic.[18]

Morality, or proper behavior, is an informal power based on the "bonding quality of personal reciprocal relationships between superiors and subordinates."[19] These differential relationships have prescribed behavior patterns which both people are expected to follow. "Chinese subordinates are still trained and encouraged to follow prescribed behavior; i.e., to conform to the orders of the authority and to show the assumed emotion (behavior) to the authority."[20] If one behaves in the appropriate manner, there is the expectation of being accepted and looked after by the larger group.[21] Because of the number and variety of relationships

---

16. Fei, 73.

17. Ibid., 24.

18. Ibid., 74.

19. Ibid., 326.

20. Chung-fang Yang, "Conformity and Defiance on Tiananmen Square: A Social Psychological Perspective", in *Culture and Politics in China* (Newark, New Jersey: Transaction Publishers, 1991), 202.

21. Pye, p.328. It is on this point that Pye prefers the term 'dependency,' whereas Fei Xiaotong uses the word 'morality.' I have chosen to use the word 'morality' because it better captures the idea of a culturally defined proper or ethical behavior. Further, because opposition and change are

the degree to which Chinese ethics and laws expand and contract depend on a particular context and how one fits into that context. The first thing to do is to understand the specific context: who is the important figure, and what kind of relationship is appropriate with that figure?[22]

Although a subordinate follows the rules of expected behavior, there are also the expected reciprocal actions on the part of the higher-level authority figure. This is described by Fei Xiaotong and Lucian Pye as paternalistic power. Paternalistic power, a term relevant to the traditionally male-centered family hierarchy, is based on education and forms the structure of power within China. It is "the transferal of knowledge from the knowledgeable to the ignorant,"[23] the passing on of experience and the trust that the knowledge passed on is in fact useful for contemporary situations. Education is important because all social rules are man-made, which lies in contrast to the notion of natural law in the West, and as such they must be learned and practiced. Dictatorial or consensual power may describe limits of individual behavior in various settings, but without learning and practice they will be ignored. "Learning requires some form of compulsion, and behind this compulsion there must be some form of power," and this power takes the form of paternalism.[24]

> Power holder's authority derives from their 'educational' function. They maintain order by upholding the morality of differential relationships in which they themselves hold a superordinate position. Because they hold a superordinate role and likely do so in multiple networks of relationships (e.g. family and village), certain power holders become responsible for everyone's proper conduct in the network. The more distant one is from the core for a network, the less one can legitimately arbitrate the conduct of

---

possible in social relations and in the definition of what is ethical, there is not an absolute situation of dependency.

22. Fei, 78.

23. Ibid., 131.

24. Ibid.

114

people in the network. Although political officials may have educational, coordinating, and even corrective roles, they are distant from local society and have circumscribed duties in relation to it. In principle, officials are responsible for managing the whole, but they do not intervene in the parts. The only exception occurs when the parts threaten the stability of the whole. When disorder occurs, all within that immediate network are at fault, and each successive circle has a duty to intervene and reestablish order, village to family, county to village, etc.[25]

Paternalism, through institutionalized networks of relationships, establishes rulers who are distant from daily affairs but serve as

role models and guides for behavior. As long as the ruler maintains the assumed emotion (behavior) toward his subjects by attempting to make his subjects' lives stable and prosperous, the ruled often are indifferent about who is ruler and happy to show at least assumed conformity to whomever it is.[26]

Paternalism and morality, as guides for behavior, serve to promote a sense of community at the local and national levels.[27]

---

25. Ibid., 29.

26. Yang, 204.

27. Ibid., 330. Minchuan Yang in "Reshaping Peasant Culture and Community: Rural Industrialization in a Chinese Village" concludes that peasant relationships are changing due to industrialization; they are not as patrilineal as they are affinal. The village community has been fragmented by rural industry into "socially and economically diversified groups," and that this "is a part of the process of emerging peasant family individualism" (p.175). Although patterns of interaction are changing and breaking traditional village boundaries, Yang does not define the concept of 'individualism' in relationship to the social structure or differentiate it from a Western definition of 'individualism.' Yang gives an example of factory workers who would rather employ hired farm workers than family to avoid the expense of a traditional feast for the family help. He cites this as an example of a reshaping of "reciprocity and interaction."
This may be true, but it is without examination of other forms of reciprocity or of reciprocity in the new structure of relations. We must ask if people's sense of family, responsibility, duty, expectations of behavior are being fundamentally altered or are outward manifestations changing to conform to a changing economic situation without a change in attitude? For example, are affinal relations taking on the same characteristics of expected reciprocal behavior as the patrilineal relations? Are ideas of family responsibility being changed or are they being extended to include relationships based on a new social situation. If the concept of 'family' is being altered to encompass affinal relations this should be considered a change in degree rather than a fundamental shift towards our measure and expectation of 'individualism.'

It is through these behavior patterns that each person within a relationship is capable of manipulating the other and expressing satisfaction or dissatisfaction.

> Opposition to what the elders say takes the form of an 'interpretation.' Such interpretations outwardly maintain the form of paternalistic power but change the content. People have had to justify social change by reinterpreting the old authorities. Such reinterpretations lead to a widening separation between names of phenomena and their reality. Under such circumstances, hypocrisy is not only unavoidable; it is a necessity.... Distorted interpretations become commonplace. Position and power, name and substance, words and deeds, theory and reality - all tend to separate and lose their interconnectedness.[28]

During periods of rapid social change there is a shift from relying on the teaching of experience, which is no longer relevant, to experimentation with theories and philosophies that can serve as guides to restructure experience.

> On the one hand, holding on to ineffective tools (methods) is senseless and inconvenient and may even result in losses. On the other hand, new methods are not ready-made but have to be invented by someone or learned from some other culture. And then the new methods have to be tested, and only afterward will people begin to accept them.[29]

The general nature of a good, or the ideal differential relationship, is one in which there is the expectation by the subordinate that the supraordinate will wisely set goals and educate the subordinate about achieving the goals. Poorly

---

Yang's conclusion is that he has found an indication of emerging individualism that will "promote market exchange" and possibly undermine China's ability to modernize and "maintain its Chinese characteristics."(p.177) This presumes that 'Chinese characteristics' must remain in a traditional stasis. If however the Chinese concept of individual' remains a function of the social relations that an individual is a part of, and if the 'like a family' metaphor continues to have meaning then there will indeed be a modern China with Chinese characteristics that is an extension of its past into the present which will be qualitatively different from Western expectations.

Minchuan Yang, "Reshaping Peasant Culture and Community: Rural Industrialization in a Chinese Village," in *Modern China* 20, no.2 (April 1994), 157-179.

28. Fei, 132-33.

29. Ibid., 129

116

established goals and/or useless education leads to distrust and usually passive opposition. As long as the central government establishes development goals that continue to make China modern, that continue to improve the quality of people's lives, and that provide people with the knowledge that allows them to see the value of new methods and then to adopt those methods, social stability and growth will be maintained. It is up to the government to establish the correct policies, this is the nature of social relations and modernization that are implied in the concept of agriculture as Chinese.[30]

As a metaphor 'ecological agriculture' serves to create a sense of a uniquely Chinese solution to a uniquely Chinese dilemma, how to feed a rapidly increasing population, on limited resources, protect the environment for future generations, and increase the income of farmers. It legitimizes the use of traditional, nonmodern methods as being both historically valid and contemporarily useful. It was the use of these methods that saw China through its difficulties in the past and it will work again now. The uniqueness of China's situation demands the use of uniquely Chinese methods. Traditional methods are modernized as concepts through scientific research that increases understanding and technological advancement through the use of biogas digesters and the calculated application of chemical additives.

---

30. Though this explanation of Chinese social relations may seem to be a step back to more traditional forms that are in question or in flux, they are still viable and fundamental for understanding contemporary China. The following quote supports the point: "A deeper level of ideology overrides the opposing political philosophies (Nationalist Party in Taiwan and the Communist Party). In that respect, the centuries old Confucian tradition of subjugating the individual (education per Xiaotong) to the institution in return for social stability has molded the monarchical, republican, and socialist polities in more or less the same fashion.... Marxism does not teach moral philosophy or formal ethics per se. In China, where politics has traditionally been perceived as an extension of moral philosophy and social adjustment is often effected through a sense of decorum rather than explicit law, this void in the imported theory of government has to be filled with what is intrinsic to the host culture. Without borrowing from Confucianism, communism could not have been operational in China. The Sinification of Marxism has occurred in the
interaction between patriarchal Confucian ethics and scientific Marxist politics, resulting in a semblance of the mandate of Heaven that the party can present to the populace." Zhang Shuqiang,"Marxism, Confucianism, and Cultural Nationalism," in *The Chinese and Their Future*, 96.

Ecological agriculture is an attempt to bridge the technical and historical gap between the two categories of traditional Chinese and Western chemical based agricultural production. The final result is intended to create a contemporary resolution that can move beyond the problems of method as they are independently applied. Through extensive research that explains why and how traditional methods are effective, by extending the principles of traditional agriculture to new plant combinations and new physical environments, and by showing how traditional methods can be improved upon with a controlled incorporation of Western petro-chemical know-how traditional agriculture becomes modern because it has become scientific.

Ecological agriculture as a metaphor emphasizes the inter-relationship between humans and their natural environment, between society and government, and China's historicity and sense of self-identity. In this way it provides the conceptual foundation for action. It is modern, it is fruitful, it is Chinese.

Communication Beyond the Metaphor's Experiential Zone

"A metaphor can serve as a vehicle for understanding a concept only by virtue of its experiential basis."[31] Domestically professional orientations gave a different interpretation as to what the underlying motivation for ecological agriculture should be, either environmental protection or agricultural production. Given the method of ecological farming as described in Chapter 3, it is apparent that both are strongly interrelated, so between agronomists and environmentalists there is a shared meaning about the principles, but the different professional experiences lead to different interpretations of what should be the primary objective.

---

31. Lakoff and Johnson, *Metaphors We Live By*, 18.

118

For the Chinese, ecological agriculture is distinctly different from the concept of sustainable agriculture in the West.[32]   In the West sustainable agriculture can mean many things, but from the Chinese point of view Western discussions of sustainable agriculture specifically refers to organic or nonchemical agriculture.[33]   Additionally in the United States it is primarily up to private organizations and individual farmers to discover and implement sustainable methods, whereas in China there is a much stronger governmental role for educating and assisting farmers.   The confusion by Americans towards the Chinese concept is in understanding the role of government and the integrated or ecological theory of ecological agriculture, the ability to see ecological agriculture as a social and agricultural philosophy.

Because of competing interpretations of meaning, the term 'sustainable agriculture' has been adopted by the Chinese for the sake of allowing dialogue between different groups both nationally and internationally.   Another example, the use of the term 'nature farming' by the Japanese is very disagreeable to the Chinese whom I spoke with.   Although the goals of nature farming are similar, that is, to develop farming methods that do not rely upon chemical inputs, the term as a metaphor fails to capture the totality of relationships that Chinese philosophy of ecological agriculture does.   From the Chinese view farming is not natural; it is inherently exploitative of nature, and as a term 'nature farming' misses this expletive and yet integrated point encompassed in the meaning of ecology (sheng tai).

Using the term sustainable allows different groups to engage in intergroup dialogue in spite of experiential dissimilarity.   For example communication

---

32. It is interesting that traditional Chinese agriculture was the substantive source of ideas for American sustainable agriculture, and when asked about the where the ideas for ecological agriculture originated I was told with a laugh, that they came from the West. The West, which primarily refers to the United States, is held as a model of modern scientific thinking and practice.

33. For a discussion of the many meanings of sustainable agriculture see Michael Redclift, *Sustainable Development: Exploring the Contradictions* (London: Routledge, 1987).

between the Chinese, the Japanese, the Koreans, and the Americans, as well as within China between agronomists and environmentalists, was facilitated by the use of this term. The ideal of sustainable agriculture as a metaphor has a more widely shared, albeit more fuzzy, experiential basis that rests upon a concern for renewing limited resources while maintaining or increasing production levels. The benefit of using a more widely accepted metaphor is to bring groups together and facilitate a discussion of methods and differences that might otherwise be overlooked if only the definitional differences were emphasized.

The 1994 establishment of the World Sustainable Agriculture Association (WSAA) branch office in Beijing and related conference is important for China's inclusion in the international sustainable agriculture movement. By adopting the term 'sustainable,' the unclear term 'Chinese ecological agriculture' is no longer a point of discussion but allows for international dialogue while maintaining a uniquely Chinese meaning. This terminology shift has allowed CSA to move from being an isolated domestic program to one that is now plugged into the international network of sustainable agriculture agencies.

Having just returned from participating in the WSAA conference in Beijing, an agronomist with whom I had spoken several times before began to tell me about the important significance of the word 'sustainable' for establishing a base of acceptable assumptions about development that allows people to overlook definitional differences and confusions and to engage in a fruitful exchange of ideas. In fact he was so taken with the idea of 'sustainable agriculture' and the relatively minimal difference between the Chinese practice of ecological agriculture and that of sustainable agriculture in the United States, that he believed ecological agriculture should be renamed as sustainable agriculture. His belief in the need for this renaming came from two concerns: that ecological agriculture be seen as only concerned with environmental protection rather than with increasing agricultural production and that, as he had been told by a professor from UC Berkeley, owing to China's great cultural and geographic

120

diversity one could not categorically say exactly Chinese agriculture is with any expectation of uniform application. So trying to define a specifically Chinese form of agriculture was unprofitable, for even ecological agriculture changed form and character with local conditions, and using the more general term of sustainable agriculture allowed broader application.

After his discussion about the importance of the term 'sustainable,' he told me several stories intended to illustrate the usefulness of the term, but his stories only conveyed to me that the original philosophy of ecological agriculture and its uniquely Chinese character remained firm.

The first story, which had also been reported in the local papers, was about the overuse of pesticides in the Nanjing area. During the summer of 1994 the weather had been particularly hot and conducive to an unusually large number of garden pests which were fond of cabbage. The farmer's tendency to distribute pesticides over their crops every time an insect was seen, rather than at appropriate times during the season, led to such high levels of chemicals that the cabbages had to be soaked in frequently changed cold water for at least two hours to leach out the pesticides.

The second story was about a sustainable agriculture project he had worked on with some Western researchers. The work being done involved a bean crop which had unfortunately become infested with aphids and was being destroyed. One of the Chinese researchers suggested that they needed to spray the bean crop in order to kill the aphids and save the beans. The Western researchers opposed this idea in favor of a more naturalistic approach. They had noticed that a nearby alfalfa field was full of aphid-loving lady bugs and suggested that once the alfalfa was harvested the lady bugs would migrate to the bean field of their own accord and eat the aphids. This idea was accepted, but proved to be fatal for both the Westerner's idea and the bean crop, for by the time the alfalfa had been harvested the aphids had destroyed the bean crop.

The point of the stories was that on one hand China's over reliance on chemical pesticides and fertilizers cannot continue for health, production, and environmental reasons. On the other hand a strictly chemical-free approach to sustainable agriculture as is used in the United States is also untenable because China must maximize production and cannot rely completely upon natural methods.[34] From this man's perspective and from others with whom I spoke, it was clear that China must find its own way by combining the technology of the West with the social and environmental conditions of China.

"Chinese ecological agriculture reflects the marriage of oriental wisdom and modern science and technology."[35] Understanding the relationship between traditional methods and Western science is a main concern of the literature on ecological agriculture. Traditional agriculture, the oriental wisdom, is described as long-term, harmonious, balanced ecologically, self-sufficient, complementary, stable, sustainable, and rational. Western science is much more unidimensional in its characterization; it is described as chemical, external, artificial control; it is a single, advanced technique that cannot meet multiple objectives; it is modern; and

---

34. This is based on the erroneous understanding that sustainable agriculture projects in the United States are primarily organic, or non-chemical, and overlooks the wide range of possible methods that are considered to be sustainable but not necessarily organic.

For everyone that I spoke with, maintaining and increasing production was the primary goal of ecological agriculture, both for convincing farmers to implement it and as the positive economic benefit of the method. As the acceptability of ecological agriculture becomes more widespread only then will the understanding and importance of environmental protection be more acceptable to the farmers by understanding through practice. Layered upon this evolutionary concept is the more futuristic, and for now very Western notion of Organic Agriculture. Organic agriculture is only now being experimented with in a newly created office, the Organic Food Development Center of the China Environmental Protection Agency, and it is intended to be the next step after the complete achievement of ecological agriculture.

35. Cheng Xu, "'Hard' and 'Soft' Restraints for China to Sustain Agricultural Development and to Follow the Conventional Modernization Approach and Deserved Alternative Way," in Shi Yuanchun and Cheng Xu, eds., *Integrated Resource Management For Sustainable Agriculture: Proceedings of the International Conference* (Beijing: Beijing Agricultural University Press, 1994), 415.

122

it is a hard technology. The intention is to combine the two methods for a "rational allocation of resources."[36] Integration will rely on

> system engineering (macro political coordination and correct policy making) to assemble various techniques (traditional and Western methods) according to local conditions in order to realize integral function and transformation of a technological system from characteristic of technical breakthrough (single use technologies) to that of systemic integration.[37]

This concern for the careful integration of Western science follows from seeing that all social situations by their very nature are ecological, or interrelated, and that there is a need to establish proper relationships. The differences between 'oriental wisdom' and 'modern science' are also seen in definition of the English 'ecological' and the Chinese 'sheng tai', which is the translation of ecological. The English definition of ecology is as the "totality or pattern of relations between organism and their environment," i.e., human ecology.[38] From 'eco' meaning habitat and 'logy' as theory or doctrine, ecology represents the Western scientific search for why and how different patterns of relationships exist.

Compare this with shengtai which is also a noun. 'Sheng' as a noun means existence or life, and 'tai' as a noun means form or condition. The Chinese definition includes the notion of relations between organisms and their environment but goes further to encompass the more important and fundamental point that these relationships are the form, the condition of life. Both words recognize the interrelatedness of being, but there is a depth in the Chinese notion, reflective of Taoist thought, that emphasizes life and the dependence of life upon the interrelatedness of all things. This understanding of interrelatedness as a

---

36. Zhang Renwu, Ji Wenying and Zhang Tong, "Analysis on the Characteristics and Components of China's Sustainable Agriculture Technological System," Paper presented at the World Sustainable Agriculture Association conference, (Beijing, 1994), 1.

37. Ibid., 2.

38. Webster's Ninth Collegate Dictionary, 395.

primary condition of being corresponds with the earlier discussions of morality in Chinese social life and supports Chinese comments calling for the integration of 'oriental wisdom' and 'Western science.'

In spite of the shift to the generic term of sustainable agriculture it is evident that the underlying philosophy of ecological agriculture, agriculture with Chinese Characteristics, is still the primary concept motivating actions. Chinese researchers and government officials that need to carry on interagency dialogues, whether domestic or international, use the term 'sustainable' for the sake of allowing discussion, but then when individuals return to their work it is with the specific organizational or professional meanings which they give to the concept. Domestically the continued use of the term ecological agriculture is prevalent; in fact when asked about the term sustainable agriculture most people were unfamiliar with it.

The term sustainable agriculture serves two purposes: it symbolizes China's ability to constructively participate in international dialogue by adopting a metaphor that goes beyond the experiential limitations of ecological agriculture. Sustainable agriculture as an outward facing metaphor is important for supporting an identity through interaction which is modern, forward looking, international and still allows for a uniquely Chinese interpretation of what sustainable agriculture means.

Implementation: The Ritual of Reciting Goals
and of Following Models

People are framed between the remembered past and the imagined future with the need to fill the inchoate present with activity. We are 'time binders' concerned to find the kind of activity that will fill this frame and bind the past and future together. This need is even more pressing in transitional societies moving between tradition and modernity. In these societies, so painfully poised upon the uncertain interface of the past and future, there are few well-proven frame filling technical and ritual routines.[39]

The implementation of the ecological agriculture project is based on two important rituals: that of achieving a goal by establishing a model and that of that of reciting goals and methods. These rituals served to actualize the ideas and goals authorized by the metaphor by connecting past to present and by clarifying actions that will carry people into the future.

The implementation of the project was itself rather simple and straightforward. Relying on word of mouth, the media, and the successes of model farms, it seemed to me to be more like the process of religious conversion based on active missionary work rather than the forced implementation that one would expect from a government-sponsored project. Once it was decided to go ahead and introduce the idea of ecological agriculture on a widespread basis, which formally occurred in 1986, the national government allocated enough funds for each province to establish an official office within the Provincial Department of Agriculture, the Office of Rural Energy and Environmental Protection. Within each province a certain number of model villages, in Jiangsu the original number was 100, and within each village only several households, those thought most likely to succeed, were selected based on that household's willingness to volunteer to serve as a model farm.

These model villages and farms were given the funds and assistance necessary to establish an integrated system of farming. This usually involved adding sheep, pigs, chickens, and/or cows to the family's resources, a biogas digester, and the seeds or seedlings necessary to diversify their crops. The whole farm then became managed as described in Chapter 3 so that animal, plant, and human waste all served to fertilize and/or generate methane, depending on the

---

39. James W. Fernandez, "The Performance of Ritual Metaphors" in J. David Sapir and J. Christopher Crocker, eds., *The Social Use of Metaphor: Essays on the Anthropology of Rhetoric* (Pittsburg: University of Pennsylvania Press, 1977), 118. Fernandez defines ritual metaphors as a devices by which new meanings are learned, and actions are generated.

local needs and conditions. After several seasons, and the obvious prosperity of the new farm in terms of increased production levels, increased family income, increased soil fertility, decreased waste, which resulted in an overall increase in the standard of living for that family the model farm was ready to serve its function as a model, or showcase, to other farmers. The success of the farm was spread verbally and through available media such as television, radio, and newspapers.

The concept of a model was prevalent in my discussions about sustainable agricultural development and economic development in general. For the Chinese I spoke with having a model provided a goal to work towards, and being a model (China as a model or one's work place as a model) gave one a great sense of purpose, social importance, and personal pride. The concept of a model was acknowledged as being important though those interviewed could not explain why.[40] Some examples that came up were the following: the United States is the model of a developed country to be achieved in terms of technology and industrial ability; CSA as a model of sustainable development within China and for other third world nations, especially African nations; China's relationship with the World Bank as a model for other third world nations; in the implementation of CSA model villages were selected and volunteers within the villages to serve as model farms.

> All societies use models in one form or another as part of their socialization process. Models are used to perpetuate traditional values and to infuse new ones, to provide social cohesion by establishing a shared body of beliefs, and to adapt society to new needs and situations. They provide social stability but may also be active agents of social change. Models provide vivid, concrete examples to the people of each society, especially their young, and it is anticipated that the values and norms represented by Models will be internalized by others an lead to emulative behavior....China has gone farther than other societies in institutionalizing the selection and publicizing of models as

---

40. These interviews included two prominent sociologist in Nanjing and in Shanghai.

vehicles of socialization and as a means of social control. It exceeds others in the degree of organization, frequency and ubiquity of use, and in the political importance given to modeling in the governing system. It is an exaggerated case of human engineering for purposes of social, political, and economic change, but also for stability once the desired change has taken place.[41]

Models, either individuals or a group, in China are selected by a formal or official selection process. They are ordinary people chosen by their peers through a carefully planned and regularized process based upon their embodiment of outstanding qualities that all people can emulate. The process ensures that the model is well known to those around him or her, and "through personal, daily, face-to-face contacts can readily communicate and explain in appropriate terms the messages and directives sent down from above."[42] Models are not extraordinary but are a part of normal, everyday life. They can be one's neighbor or work partner; it is someone with whom one has frequent contact, and as a model is a subject of discussion locally at meetings and in the press.

There is a strong historical precedent in China for the use of models as agents of social change. "The emulation of models was an important Confucianist principle....it was not just one way of learning; it was by far the most efficient way, and one could inculcate any virtuous behavior in people by presenting the right model."[43] In Chinese the word *xue*, 'to study,' originally meant to imitate, or emulate.[44] Mao also believed in the emulation of models as the best way to teach and to make people understand. Though there was a conscious adaptation of

---

41. Betty B. Burch, "Models as Agents of Change in China," in Richard W. Wilson, Amy Auerbacher Wilson, Sidney L. Greenblatt, eds., *Value Change in Chinese Society* (New York: Praeger Press, 1979), 122.

42. Ibid., 124.

43. Donald Munro, *The Concept of Man in Early China* (Stanford: Stanford University Press, 1968), 96, in Burch, 125.

44. Perry E. Link, *Evening Chats in Beijing: Probing China's Predicament* (New York: Norton Press, 1992), 10.

Lenin's theory of organization, establishing a vanguard, the Chinese rejected the emphasis on individual competition and Mao become more concerned to develop "virtue and virtuous men who act with revolutionary zeal, and with educational reforms to produce such men."[45] Mao understood that change

> does not occur spontaneously but must be motivated and directed by external sources. Change in attitude, and therefore behavior, is not self-generated by the individual himself. Once motivated, there must be an input of will and determination on the part of he individual, but motivation comes from outside.[46]

Three aspects of following a model are familiar to our Western experiences: the imitation of a model, the inspiration by a model which motivates, and the competition with a model or emulation to equal and surpass the model. A fourth aspect is emanation, and it this that explains why model emulation is so much a part of Chinese culture and behavior.

> Emanation refers to the ability of virtue (as a Taoist and Confucianist concept) in itself to directly affect, without the intervention of will on the part of the model or those exposed to him, the attitudes and behavior of others. It explains that the call is to emulate 'the spirit' of the model, not the model itself.[47]

The spirit of the model is not charisma, a magnetic, arousing form of leadership, but pure virtue, a standard of right and moral excellence. Change is motivated then by creating a vituous model which motivates others to change themselves in a similar manner. This is not the redirection of loyalty through charisma which may or may not motivate individual change but is more likely to garner support to follow a leader's initiatives. Mao referred to this virtue as spirit or as attitude, and "when found in a high-virtue model may radiate a spirit to

---

45. Burch, 128.

46. Ibid.

47. Ibid., 133.

which others respond by themselves becoming more virtuous."[48] It is a concept that is no longer a common part of Western culture but is a viable part of Chinese culture and underlies the experience of emulation.[49]

Emulation of a model is a form of compliance. Everyone knows that models are authoritative, not in having power to punish but because they have official backing. They know that emulation is expected of them, and self-preservation dictates behavior at least overtly consonant with that of the model presented to them. This compliance through voluntary or semivoluntary pressures to adopt the official line by emulation is a very important component of the Chinese style of indirect rule.[50]

This idea plays upon the idea that the individual is defined by and through social relations and so encourages group identification and the security and acceptance that comes from emulation rather than from independent action, cooperation rather than deviance. Another element of emulation is more self-serving in nature, emulation so that one may be chosen as a model. Being chosen as a model can serve as a stepping stone for social and/or political mobility by increasing chances for becoming a low-level official or by opening up opportunities for better housing, jobs, and preference for school admissions.[51]

The self-serving aspect of being or not being a model became apparent in some of my interviews and was one drawback to the overall implementation of

---

48. Ibid., 134.

49. The exception to this comment would be C.G. Jung's development of the concept of individuation, which is an effort to re-establish this inner virtue that he believes has been lost in Western society. Based on this idea of individuation, Jung believed that society can change only through changes in the individual. "If a single individual devotes himself to individuation, he frequently has a positive contagious effect on the people around him. It is as if a spark leaps from one to another. And this usually occurs when one has no intention of influencing others and often when one uses no words." M.L. von Franz,"The Process of Individuation," in Carl G. Jung, ed., *Man and His Symbols* (New York: Dell Publishing, 1968), 245.

50. Burch, 135.

51. Ibid., 136.

ecological agriculture. In one instance research monies were used to pay a 'demonstration fee' to village leaders to allow the project to be carried out in their villages. The project was not being adapted within the village, but it did provide a relatively long-term source of income from the 'fee' and a degree of prestige to the village for supporting a research project. In another area the three-year rotation of township leaders did not auger well for wanting to be a model of ecological agriculture. The desire for rapid short-term improvement superseded interest in supporting the long-term investment required of ecological agriculture. In this case being a model for ecological agriculture was less important than the prestige garnered from being a model of others forms of development.

The use of models is the method of implementation being used in the ecological agriculture project. It is a culturally effective method for the Chinese government to educate and motivate and to achieve goals by indirect, participatory means. As a goal the model of ecological agriculture establishes an ideal to work towards, and as problem solving it leaves the definition and solution of problems to the local levels for resolution and implementation. Below are some examples of my encounters with a model farm, the write up of a project that did not use models for implementation, and with an Anhui farmer involved in spreading the word about the model of ecological agriculture.

The state-owned cottonseed and breeding farm in DaFeng County is held as a national-level model for ecological production. Established in 1952, it is a goal for others to emulate and was titled as an advanced unit by the Ministry of Agriculture in 1990. The director of the Dafeng farm explained to me the model role that his farmed played.

> In the whole county this farm is the dragon's head for cotton production. The dragon is a symbol of China, and in development represents a system project. The head of the dragon is the best part of the system, and it is the most important part. By distributing

good cotton seeds throughout the country we help develop the system of ecological agriculture.[52]

He also explained that the social values were better at DaFeng because of the high quality of cottonseeds. The popularization of the high quality seeds leads to increased incomes for the county farm, higher quality cotton and textile exports lead to higher incomes for cotton farmers in general, and the decrease in disease from using higher quality seeds lowers loses for everyone, and in this the whole country benefits. Pride in producing high quality seeds, in fulfilling the function of the dragon head, is a higher social value because ultimately the whole country benefits; the perfection of the individual to benefit the group.

A Sino-German sustainable agriculture project in Shandong Province followed the same principles of ecological agriculture, but without using models for its implementation. The project began in 1990 and is a villagewide project that involves the improvement of drinking water accessibility, irrigation, soil and water conservation, reforestation, and some solar energy experimentation. "Suitable regulations" are required and are accompanied by "strict enforcement" such as forfeiture of rights to resources, or the broadcasting of violators names over a public broadcasting system.[53] When I asked an extension worker at the Jiangsu Department of Agriculture about the project's need for enforcement, he

---

52. "The Eastern dragon is not the gruesome monster of mediaeval imagination, but the genius of strength and goodness. He is the spirit of change, therefore of life itself. ... It has nine resemblances, or forms, viz.: the head of a camel, the horns of a deer, eyes of a rabbit, ears of a cow, neck of a snake, belly of a frog, scales of a carp, claws of a hawk, and palm of a tiger." C.A.S. Williams, *Outlines of Chinese Symbolism and Art Motives* (New York: Dover Publications, Inc., 1976), 132-133. I was told that these forms represent the different peoples and regions of China, symbolizing their uniqueness and unity as peoples of China. Author interview, November 1994, Dafeng, People's Republic of China.

53. Ramesh C. Agrawal, "Resource Use/Management, for Sustainable Agriculture: Some Experiences in Maguyu Project Shandong Province, China," in *Integrated Resource Managment Conference*, 275-287. Other problems with the project included introducing expensive solar energy equipment. Solar technology was beneficial, but only 10% of the homes recieved it, and the other homes can never expect to be able to afford such equipment. The cost of repairs is also prohibitive. It is also safe to assume that the village leaders received a 'demonstration fee' for the project which would have helped motivate writing the necessarily strict rules.

quickly, and adamantly responded that such measures would not be necessary if the villagers had been allowed to learn for themselves. He told me that he often has farmers coming to him that have recently decided for themselves to try the new agricultural method and that their high enthusiasm for learning and following the prescribed methods is one of the high points of his job. "The farmers want to learn, they are good people, but they are also very practical and must first see the benefit of something for themselves."[54]

My chance meeting with a farmer from Anhui province was the most clear example of an individual as a model and of how reciting the goals, methods, and benefits of practicing ecological agriculture were the most powerful tool in spreading the concept throughout the general population. It was a chance meeting that occurred one day while I was at the China National Environmental Protection Agency conducting other interviews. My host at the agency asked that if I was interested he would invite the farmer to have lunch with us, and of course I was. His excitement and enthusiasm for discussing his work were immediately apparent. He had been in animal husbandry for the last 30 years and is now a township extension worker in animal husbandry. The township has twelve villages and about 10,000 people.

Rural industry had been tried in his area, but for various reasons it had not worked and was now nonexistent. The survival of the township depended solely upon agriculture, and it was the implementation of ecological agriculture in the 1980s, motivated primarily by the excessively high cost of chemical fertilizers, that was responsible for the current growth of incomes within the township. He took obvious pride in recanting to me the technical aspects of this new way of farming and in being able to explain why, for example, stereo-cropping or intercropping work or how to set up and operate a biogas digester. For this man

---

54. Agriculture extension worker, interview by author, November 1994, Nanjing, People's Republic of China. Similar comments were made in interviews with Chinese and Western researchers, and township officials.

this was not the revival of traditional methods but a new and scientific approach to agriculture, and it was saving his township. The ability to illustrate his story with scientifically based information added a legitimacy to his authority and thereby to the new methods he was advocating.

It was through his job as an animal husbandry extension worker that he would tell other farmers of the benefits of ecological agriculture. While visiting a farm to specifically discuss pigs or cattle, he would casually bring up the topic of ecological agriculture. Have you heard about this method? Did you know that the Li family is practicing this method and their income has increased? If you are interested, I could tell you how. If the family was interested but could not afford to get started, especially with the cost of a biogas digester, new stables and sewage feeders, this extension worker told me that he lent them the money himself as a three-year loan. Nor was it just a local issue, but always began with a discussion of China's resource and population problems that need to be overcome to insure continued independent development.

This farmer exemplified to me the processes of conversion. Through the ritual of repetition he extolled the virtues of the model, inviting people to see and try it for themselves. With missionary-like zeal, he was casually introducing and spreading the idea of ecological agriculture throughout his township. It was not overt pressure, but relied on the positive, visible achievements of himself and others, the potential for adoption by anyone, and the direct connection of individuals to national and local issues that left one at least interested in hearing more. As farmers saw for themselves that the methods worked they were more willing to try them, and the understanding of the technical terms and scientific rational behind the process increased the individual farmers own sense of ability and usefulness. Further, by using his own money as an expression of his belief in the viability of ecological agriculture this extension worker seemed to emanate the spirit of virtue. For him, that all the monies he had lent out in the last five years had been paid back in full was a measure of the success of ecological agriculture in generating social and environmental wealth and his success in promulgating the idea. I was only left to guess whether or not he profited from his loans, I assume that he did, but indeed that he was paid back in full illustrates that the investments paid off for the farmers as well, the new practitioners of ecological agriculture.

It was by the repetition of the problems, methods, and goals of ecological agriculture that the extension worker from Anhui had his greatest effect, and once new converts could also repeat the ideas they felt a satisfaction in the mastery of this new and beneficial knowledge. On my initial interviews when village leaders, and county- level leaders were asked, "what are the goals of ecological agriculture?" the response was consistently the same; it is an integrated approach to farming that will protect the environment and increase production which is important because of China's historically limited natural resources and increasing population. In fact this answer was so frequently repeated and with such consistency that I stopped asking the question, and the student who accompanied me to the countryside also advised that I stop asking the question because I would only keep getting the same answer. It is this repetition of these Chinese characteristics that are a part of the ecological agriculture metaphor discussed earlier that connects the individual historically and culturally to identity of being Chinese.

My initial reaction to this was that somehow I was being collectively lied to and that I needed to find out what was 'really' going on. I believe that this initial reaction was the product of my own initial assumptions: (1) that as an outsider I would only be given the politically acceptable line and (2) that my own biases and distrusts partly of institutional answers, but surprisingly I felt a certain distrust because I was speaking with communists, a product of my own culture. As discussed in Chapter 2, I had to shift away from asking my list of questions generated prior to my arrival in China and focus on points and issues raised during conversations. My questions served as guides for flagging issues to question more closely as they emerged in the conversation. The result was to learn more about what different individual believed was important about the project and their work with it.

As my research continued it became apparent that for many people the initial response did reflect what was going on in terms of the problems to be addressed and the belief in the possibilities of the model of ecological agriculture. For others, though, there was no general disagreement with the statements of problems and goals. There was doubt as to the ability of the project to deal with the enormity and urgency of China's problems and the feeling that the project itself largely served immediate interests. For example, it was seen as only a way to keep researchers busy, or as simply the renaming of traditional methods to keep the farmers motivated as will be discussed more thoroughly in the next section.

Repetition is a form of education, a form of story telling that connects the ideas of the ecological agriculture metaphor and the practical acts of the model with general population. Following a model is in itself a form of repetition in that the spirit of the model is learned through observation and listening and then applied, and as more people learn of the model there is a ripple effect out from original models, then newer models as it becomes more popularized. The term ecological agriculture evoked a sense of purpose, or mission among many that I spoke with. The ritual of reciting established and clarified the problems, the goals to be achieved, and a method in ecological agriculture for achieving those goals. By verbalizing the components of the metaphor, ecological agriculture or agriculture with Chinese characteristics, there was a reenforcement of their Chinese historical roots which allowed each individual to see themselves as directly contributing to the solution of one of the nations more significant problems. It established a final objective for people to work towards that is clean, productive, lucrative, and forward thinking in its sophisticated integration of resources. Repetition of the goals keeps farmers, researchers, and implementers motivated and feeling that forward momentum is being maintained.

The central government is expected to allocate funds for the establishment of more model villages in the next five-year plan, and as word of the success of those farmers already participating spreads, more farmers want to learn how to apply ecological farming methods. When one considers that the initial 1986 investment in Jiangsu province was for 100 model villages, of the 2000 in the province and within each village only about 10 homes volunteered to participate (participation is voluntary), and finally that wary onlookers will watch the models for one or two seasons before making a decision to implement on their farm progress is indeed very slow.

Conflicting Goals: Ecological Agriculture as a Myth

The rituals of reciting and model following generate an image of progress and action which generate hope for increased prosperity individually, and equitable growth among social groups through agricultural development that works within a culturally understood concept of social relations. The enactment of these rituals gives affirmation to the myths of the priority of agriculture in national development and the possibility of agriculture for personal gains. The mythic nature of ecological agriculture emerges when the project is viewed as a whole and is contrasted with existing realities surrounding its implementation.

This fuels a myth which I am calling the priority of agriculture that acts to, at least temporarily, resolve the tension between increasing rural and urban income disparities, and the central government's continued drain of financial resources from agriculture to industry and urban development. By continuing to emphasize the priority of agriculture in the face of its diminishing prosperity and social status the intention is to keep the majority of the population motivated, involved, and benefiting from the national modernization process. It is an effort to maintain social stability during a period of increased inequality and urban/industrial development that is legitimate on its own grounds as a goal as much as it is symbolic of a majority of the Chinese population. The official or more strategic political purposes of increasing production and environmental protection are intended to give farmers an opportunity to earn more money and close the income gap between rural and urban areas, and among those who have opportunities from rural industry and those who do not. It is also intended to be seen as an incentive, through its money earning potential, to stem the flow of farmers from rural to urban areas. This is in line with my earlier definition of a myth as being a socially constructed account of exemplary behavior and significant events in the life of the polity that provide a way to understand action and events and can temporarily hold competing values in tension by rationalizing anxiety and directing attention elsewhere.

The following indicates from an economic perspective the degree of 'priority' that agriculture is given. In 1979 the government budget expenditure on agriculture was 11.4% of the total, but by 1985 had declined to 3.6%. In 1990 this figured increased to 5.3% and in 1995 to 7%, however when this is adjusted for inflation the increase is negligible. In 1990 the net agricultural expenditure was 18.15 billion yuans, but only 8.91 billion when adjusted for inflation.[55] Income outflows occur through low-price procurement of major farm products, countered by high-cost inputs for farmer, decreased budget expenditures, the banking system, transfers from urban to rural areas and from farming to township and village enterprises have all served to diminish money available for agricultural investment.

> In 1992 the year-end outstanding of farmer's total savings in the banking system was 286.7 billion yuans, 76 billion were lent to farmers, 117 billion were used by rural industry, and the rest, about 98 billion were transferred to urban areas. As a result the total income and capital outflows from the agricultural sector increased from about 30 billion yuans in the early 1980's to more than 100 billion in the late 1980's annually. The total transfers in 1990 was roughly equivalent to 20% of GDP generated in the agriculture sector, 35% of the farmer household's fixed productive assets

---

55. Deflated by retail price index, at 1979 price level. Source: Calculated from data contained in the *Statistical Year Book of China* (Oxford: Oxford University Pressa, 1994).

measured in original values, and more than 10 times farmer's annual investment in productive assets.[56]

The large capital outflow is due to government policy that places economic priority on industrial development and maintains enough agricultural input to support food needs for the population and capital gains for industry. "An average rural household pays 2% of its income as net tax, whereas an average urban household receives a net subsidy of nearly two-fifths of its income."[57]

Everyone I spoke with (officials, students, researchers, farmers) acknowledged the importance of maintaining social stability for China's continued growth and development. Even among those who believed that there needed to be fundamental changes in the current approach to agricultural development, they believed it should not be at the expense of stability. I was several times reminded that political chaos in China has historically been violent, with great loss of life, and everyone over the age of thirty has survived at least one period of extreme social disorder. The issue of stability is not just one of keeping the Communist Party in power; it is one of maintaining a productive social order necessary for achieving China's national goals.

One night in a small cafe that served very good food, I was joined by a professor who had recently submitted a group research proposal to be included in the next Five Year Plan. I was explaining to him my interests in Chinese politics, and he interrupted me saying, "You have to understand that all of the Central Government's policies are based on two concerns: increasing production and maintaining social stability." All policy making becomes something of a juggling act between achieving the desired goals by making changes and keeping social stability among the 1.4 billion people who react in one form or another to those

---

56. Ibid.

57. Azizur Rahman Khan, Keith Griffin, Carl Riskin, Zhao Renwei, "Household Income and Its Distribution in China," a working paper in econmics, Department of Economics, University of California, Riverside, December 1991, 76.

138

changes. For example, modernization of state industries through mechanization cannot be too quick or too many people will suddenly be out of work, and care must be taken in the urban areas especially, because of the higher levels of existing organization and communication. The situation in Eastern Europe and in Russia has been carefully watched and cannot be allowed to happen within China.

The myth of the priority of agriculture expressed through CSA shifts the discussion away from regional and social economic disparities to one of progress and possibility and in doing so plays a role in maintaining social stability by providing a potential for increased individual profits from agriculture. One of the selling points of ecological agriculture is that it is easier to farm once the proper integration is in place, yet overall the method is more labor intensive even after it has been established. The method is referred to as 'meticulous farming' and owing to the close integration of plants and animals requires working by hand, rather than with machines to maintain the intense level of resource use. The potential for increased income through diversified agricultural production is off-set by the intensity of labor required. Stabilizing the increasing numbers of unemployed and the increasing numbers of people moving to urban areas could be addressed by remotivating people to do agricultural work.

The focus on the final goal, on following the model, and on what can be achieved also diverts attention away from how to achieve that goal on a wider basis. The ritual of reciting supports a second myth which I call the myth of possibility. The myth of possibility is related to the myth of agricultural priority in that it is concerned with the tension between those with income primarily from industry (nationally still a minority) and those who still rely primarily on agriculture for income. The myth of possibility shifts the locus of responsibility to the individual by providing a model for successful, profitable farming. The possibility for success is theoretically available for everyone but is based on individual farmers having sufficient funds for initial investments in seeds, technology, and livestock. Investment in livestock, such as pigs, is a slow but lucrative way to generate capital for investment in bio-gas digesters, new pens, and cash crop seeds (discussed in Chapter 3). Given the need for initial investments, currently most of the successful CSA farms in Jiangsu Province are the ones which have their primary income from rural industry and have easy access to larger metropolitan markets such as Shanghai and Nanjing.

The myth of possibility avoids discussion about how to transfer methods and technologies to villages that do not share the same economic advantages, or even within the same villages to farmers who do not have the financial resources. This is entirely left up to the ingenuity of the farmers. On one hand this is good because it permits farmers to adapt CSA technology to their own local conditions and preferences. On the other hand, how to spread the idea and how to help farmers come up with innovative ideas for adaptation are seen as someone else's problem and there is no political incentive for local leaders to invest in such a long-term, slow growth method. For researchers, emphasis is on how to continue to improve CSA as a model of agriculture to be achieved, and as one researcher told me when asked how can CSA be spread to areas that are having economic difficulties (areas that need it the most) his reply was, "that's their problem."

This seemed important to me because of the fact that such emphasis was placed on the economic and environmental benefits of CSA, and the goal of its eventual extensive spread beyond the model villages. How could the problems of diffusion and adaptation not be discussed, especially by researchers who appear to be so intent on the improvement and advancement of the concept? I believe one reason is that because of the great lack of money by all levels of government (it is reported that some local governments must pay in cigarettes and alcohol for the employees to sell for cash), a discussion of how to spread the idea would lead to a more direct and open discussions about the central government's lack of investment in agriculture and its heavy reinvestment of agricultural profits into industry, the myth of priority. This is not only a production issue, but one of social stability in light of continued reassurances about the important role of agriculture and not wanting to leave the peasants out of the benefits of modernization.

Another reason for a lack of discussion about diffusion is more cultural, or less specifically linked to the contents of the policy; it is an approach to problem solving and to the future that is significantly different from my experiences as an American. It is a conception of time that underlines the myth of possibility and is supported by the ritual of model following. Continual self-improvement of the model, whether it is an individual or a project, is a necessary process for increasing the value and effectiveness of the model. As the virtue increases, the emanating qualities of the model increase which stimulates emulation to an even higher degree.

A second aspect of following a model is that a time gap is created between the lived present and the ideal to achieve. Problems are defined and overcome in relationship to their immediacy in achieving goals. Discussion of potential problems is not largely relevant because they are only potentialities and do not at present interfere with the achievement of the goal. Further, if they do occur they would have occurred in spite of discussion and can be resolved when they do in fact create a problem. By shifting the focus from potential problems and seemingly insurmountable obstacles to issues that can be dealt with, or not, on a daily basis, the model remains a possibility and a priority and the student works in the present to achieve the goal.

While in China it became apparent to me that problems that do not immediately and obviously impede one's work are either not problems or are somebody else's problem. This attitude was evident not only in discussions about CSA, but in a variety of everyday experiences. This led me to believe that in any given task what I would see as problems or issues, where nonissues or nonproblems for the Chinese I spoke with.

My own personal experience with this attitude occurred while I was living at the guest house on Nanjing Agricultural University. Because I was a student on limited funds, it had been arranged for me to stay in the guest house normally used by Chinese visitors and not the more expensive hotellike guest house usually reserved for foreign guests and higher ranking Chinese visitors. Normally there was running hot water only from 7 to 10 p.m. each day; however for a period of time I was the only person living in this building and it was not considered economical to fill the entire building with hot water each night. So during this time I was politely told that I must shower in the employee showers out in the boiler room not too far from my building.

It was after a week of being escorted to and from my shower that I learned it was not because of a broken boiler as I originally had been told, but because I was the only guest. Upon learning this I felt put out, for though it may be uneconomical for them, I was still a paying guest, paying for a room that came with a shower. When I mentioned to the manager that it was inconvenient not to be able to shower in my room and that I should be able to use the facilities that I was paying for he replied by saying, "You are a mature man, surely you can understand that if I resolve this problem, another one will arise. Certainly, being well educated, you must be able to see this." Following a Chinese habit of avoiding confrontation with face to face encounters, I mentioned in passing to a subordinate shift manager that I would refuse to pay full price for my room if I could not use the shower that I was now paying for. That evening the 'problem' was resolved, and hot water was once again available in my room every night from 7 to 10 pm.

The idea of what was a problem and to whom had shifted. At first the problem was only mine; I was a temporary guest that did not interfere with their daily operations and so the water was not their problem to deal with; it was mine particularly since they were providing me with an alternative. The problem did become theirs when I brought up the logical issue of not paying for an unusable amenity.

Rather than maintaining an intransigent position as could have easily been done, the locus of who was responsible for dealing with the problem shifted from me back to the manager. I saw similar situations occur among the Chinese employees in everyday operations of the guest house where maintenance problems or housekeeping problems were not dealt with until normal operations were affected. A deep sink in the kitchen remained clogged for so long that the stagnant water in it had turned black and septic, but there were still three other sinks that worked. In another instance an Australian English teacher had grown so disgusted with the dirty hallway floor outside of his door that he took it upon himself to scrub the tiled area just in front of his door. To everyone's amazement the light gray gave way to coral pink tile with black boarders. About a week later, since the contrast was too much to be overlooked the problem had shifted from the Australian's sense of cleanliness to the manager's need for uniform appearance and to deal with a now more obviously dirty floor, and the entire first floor hall and foyer was scrubbed down.

For me these examples illustrate a sense of time that is very present oriented and an approach to problem definition and resolution that is very practical and realistic (not couched in the hypothetical); now it is not a problem because it does not interfere with work and later it may become a problem because it does. Discussions about the future, especially when put in the context of current trends or problems, led to answers which either obviated a discussion of the future or which led to a reiteration of the goals which were to be achieved. For example when I asked one agronomist about projected population growth and the projected drop in grain resources to support this growth, I was told to "wait and see what happens in the year 2000." One student in a very eloquent response to one of my 'what if' questions told me, "the future is like driving towards a range of mountains. Though from a distance you can not see the way through, the path becomes apparent only as you proceed." Again this is much different from my experiences as an American where future predictions are often made based on current trends, and being able to discuss anticipated problems with possible resolutions is an integral part of the planning process. In fact the number of times that it came up made me aware of how frequently I asked future oriented questions without thinking about it. From my discussions in China it seemed that planning and the future consisted of a detailed discussion of the goal to be achieved, with problem definition and resolution left to a more daily and local level.[58]

The myths of priority and of possibility reflect a temporary resolution of the tension between industrial and agricultural development by offering an alternative goal through the reinvention of farming techniques. The discussion of CSA methods and goals do not allow for an explicit discussion of the central government's economic investment in agriculture, nor for a discussion of how to facilitate the spread of CSA to areas that may be particularly needy. There is interest in trying to convince the central government of the benefits of slowing down the pace of development and to make important investments in agriculture to ensure long term growth and stability, but this is not a widespread or at this time an open discussion. Further those who expressed this view were skeptical that this point of view will be really acknowledged by central government authorities.

CSA as a myth emphasizes the priority of agricultural production that at least temporarily averts open public discussion of these larger development issues. At the same time its implementation through traditional and still socially acceptable means (models and repetition) creates the possibility for individual improvement. In shifting attention from the central government to the power of individuals in shaping their own future and simultaneously addressing the development issues as concerns of their village and country, the importance of the individual and the possibility of grass roots based improvement are accentuated.

---

58. My reactions to this comment and the frequency of this type of response were supported in my discussions with Westerners who had been living in China for a number of years.

This reflects the importance of maintaining social stability by stimulating farmer motivation through discussion of a program of possibility, and an approach to problem solving that establishes a goal without an explicit discussion of the economic and political problems in achieving the goal.

Holding It All Together

What has this meaning-based analysis of Chinese Sustainable Agriculture revealed? Through Ecological Agriculture as a metaphor one sees the strong connection and, in fact, the continuity of past into present. The connection is made by establishing a social identity that places the method and those who practice it at the heart of China's agricultural dilemma which is historically conditioned and presently real. The metaphor defines action by calling for a modernization of traditional practices that can resolve present problems by combining "oriental wisdom and Western science" to create a more productive future based on a superior and uniquely Chinese solution.

Ideally it is the responsibility of government leaders at all levels to establish good policies and to motivate people to learn new methods and technologies for achieving those policies. The relationship between people and the government is 'like a family.' The government as 'family head' has a responsibility for macro-level coordination and planning. The need for comprehensive management and the responsibility of the government to establish 'correct' policies, those that maintain unity and acknowledge diversity, is an important aspect of agriculture as 'Chinese' and it is from this basis that the form of the action legitimized through the metaphor becomes apparent; a revalidation of traditional experiences now acceptable as modern and implemented through the socially acceptable method of models and repetition.

In a society where self-identity is understood through dyadic relations, and the structure of society is based upon differential relations the use of these ritual forms also provides the channels for political power, conformity and opposition.

These ritual acts are the forms through which expected behavior is established and the process of social and political education is maintained. In this case the lesson that is being taught is that agriculture can be profitable environmentally and economically. If the individual farmers turn their attention away from the lure of urban riches, focus on what they know, and follow this scientific model of modern agriculture then they too can become wealthy and fulfill their social responsibility by being a part of the solution to China's problems.

The examination of ecological agriculture as a myth allows us to examine the conflicting realities of conformity (following the ecological agriculture model) contrasted with its problems and relationship to larger development goals (short-term political and research benefit, transfer of profits and labor to industry and urban areas) as two aspects supported by the same policy. Ecological agriculture symbolizes the priority of agriculture in Chinese national development and the possibility of agriculture for individual gains. By emphasizing historical precedent, technical modernity, and increased production, the program as a whole diverts attention away from the unchanging economic role of agriculture as a source of investment revenue for industrial development.

As a political act CSA is timely because it is a labor intensive program that for now offers an alternative way to stabilized a large percentage of the population who has become discontent with agriculture as a primary way of life and who are rapidly moving into the urban areas where there is limited or no housing and high unemployment. I do not know whether or not the government will encourage more rapid implementation with increased funds and support or allow the project to gradually grow under its own power as it now is. As a method it can generate more income to support government subsidization of urban development.

Lastly, it meets an increasing domestic awareness, that is also fueled by international pressure, to address the serious and worsening issues of environmental degradation and the diminishing resource base.

Ecological agriculture is an effective model for the integration of plants, animals and people that allows for the optimal use of and regeneration of limited resources. It is a positive model of development that has been successful as a technical program in generating interest by other farmers in those areas where it has been applied. When contrasted with the obstacles that surround the project's successful diffusion and the local conditions that have facilitated its success, the examination of the project as a political symbol allows us to integrate both positive and negative aspects for a more holistic analysis.

An understanding of how ecological agriculture as a symbolic act simultaneously supports conflicting realities is founded upon cultural themes which accommodate diverse interpretations of meaning; ecological agriculture as a metaphor, repetition, and model following (a form of repetition) as the physical enactments of the concepts of the metaphor, and as a myth in which government tries to maintain social stability without altering its methods and goals for industrial development.

The cultural themes revealed by this analysis of a contemporary policy issue are an important part of understanding the meaning that the Chinese give to themselves and their interaction with others. From this case study I believe that the following of models is a fundamental concept that has an important place in this policy's implementation and in social role taking and identity formation. Within the Chinese context as described in this chapter, model-following, as I will call it, is a function of realizing and accepting the differential nature of social relations and of the responsibilities people have in those relationships. Filling the role of a model requires continual self-improvement to maintain the status of a model and to emanate the virtues expected by those following. Following a model also requires continual self-improvement to achieve and eventually surpass the standards of the model.

From following a model it can be seen that although the goal to be realized is clear, the path towards that goal is not. Solving the problems encountered along the way are taken on a 'as they come' basis; projections are not made, trends are not marked, and problems are defined by their immediate impact on work towards the goal. In this way the goal is always realizable; problems are not insurmountable or the end itself unattainable. Following a model does not mean exact replication but adaptation of the basic elements which make a model good and applying those to one's own social, economic, and physical conditions. In this way the virtues of the model are realized and improved upon through individual flexibility and adaptability. This gives a context for considering the notion of cooperation.

Cooperation consists of each party, the model and the followers, fulfilling their roles of self-improvement and individual adaptation. Each supports the other by validating the experience created by perfecting and following the model which strengthens the process of growth and development--in a word praxis. If the model is not good, or only partially good (goodness being defined by its relevance to individual social experiences and needs), cooperation loses its meaning. In this situation cooperation changes to forms of outward deference and compliance with expected social responsibilities but does not lead to mutual support and validation. Action on the part of the follower becomes more individualistic, and although learning from the model may still occur, it is limited. Because the individual is defined by social relationships, the follower will want a new model or to become the model itself, a model which will be more relevant to existing social conditions and needs.

Models can symbolize social tensions as well as divert attention away from points of tension. The model, if it has failed to fill its obligations of self-improvement, not realizing that it has reached a point of limited or outward compliance, will be too focused on its past status to see what changes are taking place with the followers. That is, the individual or institution that is filling the role of a model can fail to keep pace with changes in social relationships that may have once defined its importance and are now less significant. The loss of trust in the current Chinese communist leadership can be taken as an example of a model that has been slow to adapt to changing social conditions. On the other hand, and to the model's advantage, focusing attention on following the model can also divert attention away from actions that are masked by the model as I have discussed in the case of ecological agriculture through the myths of possibility and potential. By focusing on the goals and values of the model as a means to overcome existing dilemmas, i.e., low farmer income, other political factors such as lack of government investment are given less consideration or ignored because they are presently seen as insurmountable problems.

By looking at this agriculture project as a symbolic act to which meanings are given and taken, one can see historically shaped identity, ritual acts, and political power expressed in contemporary forms that culturally transcend the influence of the Communist Party in China. This is significant for two reasons: (1) There is a tendency to dismiss government statements as hollow rhetoric or half truths without examining the symbolic meanings or cultural relevance of those statements, and (2) if one accepts that the CSA project reflects one aspect of contemporary Chinese understandings about social structure and political power as discussed above then one must ask to what degree do these understandings carry over into other aspects of Chinese political life.

The implementation of the ecological agriculture project through the establishment and following of models, and the repetition of the problems and goals which are summarized in the concept of ecological agriculture (agriculture with Chinese characteristics) as a metaphor reflect contemporary, everyday processes of interaction. As an agricultural policy it addresses real issues with a real and viable solution, and its implementation is a commonplace process which is culturally meaningful and politically powerful. As a symbolic act it reveals a pattern of sociocultural understanding and interaction that cannot be discounted or ignored.

CHAPTER 5

DOMESTIC ANALYSIS AND THE SYMBOLIC IMPLICATIONS
OF ECOLOGICAL AGRICULTURE

The purpose of this chapter is to consider the policy of Ecological Agriculture as symbolizing several of the larger social conflicts facing China today and as an idea that reflects one possible political and economic future for rural development in China. The issues of urban/rural, Western/Chinese, modern/traditional and how they are discussed through this chapter's analysis of the implementation of CEA offers alternative insights into Chinese political culture. I believe that these insights represent a perspective that has relevance for our American analysis of Chinese national politics, which in turn affects our understanding of international relations. This chapter will offer one possible interpretation of Chinese culture, but the important point that carries this interpretation through to the end is the need to question the supposed universality of our analytical concepts by considering what meaning may be given to them by other cultural perspectives. The result can be a broadening of our own understanding and analytical abilities.

In this chapter I will illustrate an aspect of Chinese political culture and identity that is currently active in the implementation of ecological agriculture, and when considered as a conceptual component to our analysis of Chinese politics it is one that can add depth to existing observations by forcing us to ask different questions. To do this I will build upon the analysis presented in Chapter 4 by showing how the policy of CEA is not only a technical program but also

150

reflects or symbolizes a larger social vision. Using written materials from several conferences in China, I examine my analysis of the policy as a whole to illustrate its symbolic nature. Having delineated major themes and processes that are fundamental to CEA and symbolic of key cultural themes and summarizing the conclusions of Chapter 4 and this chapter, the next step is to extend these concepts to American discussions of Chinese domestic political development. Specifically, I want to show that often when one speaks of Chinese political behavior as ironic, or contradictory, or deceptive, that an inclusion of the cultural themes highlighted by my analysis can offer a perspective that is less perplexing. Further, by reconsidering one's own analysis one is forced to become more culturally self-reflective about the concepts one uses.

Ecological Agriculture as a Symbol of Social Values

The analysis of symbolic politics is about understanding the multiple meanings that concepts have for different groups and how political acts cue the interpretation of meanings that generate support for both change and continuity. Domestically ecological agriculture is a national effort to improve human and environmental resource conditions. As discussed in the last chapter the rituals of model-following and of repetition are methods of implementation that are culturally acceptable to the Chinese, and in conjunction with the research, technical and financial support offered by the national government for this nationwide policy establish the program as a real, operational, and viable solution, but it is also a low maintenance, low investment, slow growth solution that strives to meet the immediate needs of improving farmer morale, income, and satisfaction while allowing for the continuation of an existing development policy that supports national government goals. The task at hand is to understand, through interpretation, what the Chinese involved in this project are telling themselves and others about their identity as a polity, about the relationship of power and meaning among themselves and in communication with others.

"Symbolic or discursive forms ... are constructions which typically represent something, refer to something, say something about something. It is this referential aspect that we seek to grasp in the process of interpretation."[1]

It was difficult for me to read the Chinese sustainable agriculture literature and not also to see it as an idealized model for society, not just as an agricultural project removed from other social and economic issues. In symbolizing ideal values and social structures Chinese studies of ecological agriculture (or sustainable agriculture as it is more generically referred to) reflects key values and ideas that still have meaning and significance for many Chinese. As outlined in Chapter 2, I have analyzed a number of articles using a structuralist interpretation of the narratives. In the following section I will present my analysis and show how these general discussions of sustainable agriculture support my specific analysis of ecological agriculture and present a larger sense of social relations and development. I will first present the primary sets of characters as protagonists and antagonists, then show how their relationships are described, and then lastly show what transformations are expected. This form of analysis is a discursive way to illustrate desired values. It shows who or what is perceived as the agent of change to bring those desired sets of values into prominence. In this way CEA comes to articulate values that can be conceptualized apart from the project and seen as cultural themes with political and social resonance necessary to generate action and as a concept that can be appropriated for our own understanding. It is from this appropriation that we can ask how our own concepts and questions are affected.

The key oppositions, or characters in the professional articles I treated as stories, are China/United States, tradition/technology, conventional methods/deserved alternative way, efficient agricultural development/inefficient

---

1. John B. Thompson, *Ideology and Modern Culture: Critical Social Theory in the Era of Mass Communication* (Stanford: Stanford University Press, 1990), 289-90.

152

agricultural development, economic markets/ecological environment.[2] There is a degree of overlap between these sets of oppositions that became apparent in the description of the values each represents, and even a cursory examination of the list above gives one the general idea of the issues at play. I will deal with three: (1) China/United States, which is symbolic of China/West relations and of relationships between individuals, (2) efficient/inefficient is reflective of the results of interactions between individuals, and (3) economic markets/ecological environment, which symbolizes the nature of domestic relationships between markets, people, and government. I will first present the descriptions of these relationships, then provide my analysis, and lastly place what has been learned

---

2. Chen Jiefu, Dongye Guangliang, Feng Yongjun, "Developing Sustainable Agriculture to Promote Agricultural Modernization," Shandong Agricultural University, Taian, Shandong, PRC; Cheng Xu, "'Hard' and 'Soft' Restraints for China to Sustain Agricultural Development and to Follow the Conventional Modernization Approach of Agriculture and Deserved Alternative Way," paper presented at the 1993 Integrated Resource Management for Sustainable Agriculture conference in Beijing; Ke Jianguo, Li Jincuo, "Chinese Sustainable Agriculture and Traditional Agricultural Systems," Nanjing Agricultural University, Jiangsu, PRC; Lan Jusheng, Jia Rujiang, and Wei Jiankun, "The Concept and Strategies of Establishing Sustainable Agriculture in North China Lowland Plain," paper presented at the 1993 Integrated Resource Management for Sustainable Agriculture conference; Liu Sihua, "An Approach to the Problems concerning Environment Protection and Ecology Construction in Socialist Market Economy, in *Ecological Economics (Sheng Tai Jing Ji)*, no.1 (Feb. 1994), 1-14; Liu Shukai, "The Theme of Chinese Modern Ecological Agriculture," Nanjing Agricultural University, Jiangsu, PRC; Sun Dunli, Li Chaohai, Ma Xinming, and Wu Dafu, "Improving Eco-Environment and Enhancing Agricultural Sustaining Development," paper presented at the 1993 Integrated Resource Management for Sustainable Agriculture conference; Wang Zhengping, Li Jingling, "The Key of Sustainable Agriculture is to Develop Efficient Type Agriculture," paper presented at the 1993 Integrated Resource Management for Sustainable Agriculture conference; Wang Zhaoqian, "Historical Cultural and Sociological Backgrounds of Chinese Ecological Agriculture," paper presented at the September 1995 World Sustainable Agriculture Association in Beijing; Yang Wenjing, "Study on Eco-economic Management Policy under Market Economy System," in *Ecological Economics (Sheng Tai Jing Ji)*, no.2 (March 1994), 10-16; Zhang Renwu, Ji Wenying, Zhang Tong, "Analysis on the Characteristics and Components of China's Sustainable Agricultural Technological System," paper presented at the 1993 Integrated Resource Management for Sustainable Agriculture conference; Zhang Taolin, Zhao Qiquo, Zhang Bin, "On Sustainable Agricultural Development in South China," paper presented at the 1993 Integrated Resource Management for Sustainable Agriculture conference; Zhang Xigu, Ke Jiangu, "The Key to the Breakthrough for Chinese Sustainable Agriculture: Organic Combination of Chinese Traditional and Modern Agricultural Techniques," Nanjing Agricultural University, Jiangsu, PRC; Zhou Shengkun, "A Farmer First Paradigm in Order to Complement Transfer of Technology Tendency Towards Sustainable Agriculture in China in the Nineties," paper presented at the 1993 Integrated Management Resource for Sustainable Agriculture conference.

into the context of cultural themes, making an interactive comparison between what has been learned here and other American analyses from the comparative politics literature.

Analysis of the Ecological Agriculture Literature

How are China and the United States described? The United States represents the industrialized, developed countries and is a model that has been and is being followed. At times this is explicitly stated, but more frequently it is implied by the Chinese authors in their articles through the sole use of the United States as an illustrative example for points being made. The following comments from one of these articles exemplifies this point.

> It is well known that the conventional agro-modernization approach (used by developed countries) implies reaching higher productivity by means of increasing non-renewable resources input and huge investment capitals based on scaled management and the advance of agricultural science and technology.[3]

In addition to the positive effects which established the West as a model to follow, the negative effects of this approach are also acknowledge by the author.

> Taking the U.S. as an example: its annual consumption of commodity energy on a per capita basis has been as high as 10 metric tons of standard coal, and the full process of food production covers one sixth of the nation's total energy consumption! The negative effects have become more and more apparent since the sixties. Thus every advanced country are [*sic*] searching for the sustainable agriculture approach without exception, to tackle these shortcomings coupled with high input/high output model [*sic*].[4]

The United States is an 'advanced,' 'industrialized,' 'developed' country that values the 'advance of science and technology,' can afford 'huge investment

---

3. Cheng Xu, 408.

4. Ibid., 408.

154

capital,' operates on an economy of 'scale,' and is now trying to rectify its environmental problems.

> On the other hand, "new China had been developed on a terrible poor basis of economy, ecology and environment after all [*sic*]. The resource endowment was very miserable, due to the over-population for as long as three hundred years, also, the eco-system and natural environment were badly ruined by uninterrupted war chaos and blind and depleted exploitation [*sic*]. Thus it is impossible and is not obligated for China to secure the favorable resource conditions, such as very cheap imported petroleum which advanced countries benefited during their process of agro-modernization [*sic*]. Inversely China has long been inferior in terms of per capita amount of resources necessary to agriculture even if compares with the world average level, not to mention the comparison with advanced countries.[5]

According to this story, China has been dealt an inferior hand by their own history and by their geography. It has been developing on a 'poor' political, social, and natural resource base. China has lacked the 'favorable resource conditions' of the advanced countries and is 'inferior' on a per capita measure of resources. The dichotomy established is between advanced, developed countries that have had history, money, and natural resources on their side and poor, less advantaged countries, namely, China, that are trying to develop. Due to China's relative isolation from the West after the revolution and until 1978 the agro-scientists and technicians were not aware of the negative effects of following this approach. "They keep regarding these countries as successful models of agricultural development."[6] China 'unwittingly' followed this approach, unaware that energy consumption, water consumption, pollution, land use, and lack of investment funds would be so great and pose the problems that they have. The goal for China is to develop into an advanced country for the sake of international power and domestic advancement, but the arguments made above show an

---

5. Ibid.

6. Ibid.

awareness that the model China chooses to follow can no longer be only that of already developed countries. There is the recognition by the promoters of ecological agriculture that the negative environmental effects being experienced in developed countries and their historical resource advantages are conditions that China either does not want or has not been afforded.

This dichotomy between China's situation and that of developed countries, namely, the United States, establishes the relationship between China and the United States in terms of goals and models. The United States representing an advanced, industrialized country is the goal China wants to achieve and following the approaches, that is seeing the United States as a model to learn from has been the method of agricultural development in China since the 1960s. China's increased use of high-yield varieties of seed, the chemical fertilizers and pesticides, and irrigation requirements since the 1960s are "obviously the result of following the conventional agro-modernization approach that has been gone through by developed countries."[7] China is pictured as an underdog that wants to be an advanced country because it has been and is following the model of the West, but to achieve this goal the authors of ecological agriculture argue that China cannot blindly follow this model of development. China has historically been able to overcome limitations, using labor to make up for resource deficiencies. To work with unchanging 'hard' constraints of limited resources, high population, and limited financial resources, the 'soft' constraints of attitudes towards development must change.

This leads to the second set of oppositions, efficient and inefficient agriculture. Efficient agriculture is pictured as an 'internal drive' that allows it to be 'self-developing,' or agriculture that is productive, profitable, resource conscious, and increases employment for the rural work force.[8] This form of

---

7. Ibid.

8. Wang Zhengping and Li Jingling, 574.

156

agriculture would be "high yield, good quality, and high efficiency agriculture."[9] Inefficient agriculture is represented as a 'deleting type of agriculture' that requires extensive management, exists at a low-level of exploitation, and is low-input. This current understanding of Chinese agriculture can be said to be currently deleting farmer income, agricultural supply, and farmer motivation for farming, and the natural resource base. Based upon this set of oppositions Chinese agriculture is described as currently being inefficient, and the goal is to become efficient. Efficiency is further defined by outlining its functional components:

> 1) under the guidance of market demands, following commodity economic laws to develop agricultural production, 2) relying on science and technology to optimize the combination of the key elements of the productive forces, and 3) adjusting industrial and product structures, sufficiently and rationally exploiting and utilizing natural and social resources.[10]

How, according to these authors, is efficiency achieved? Efficient agriculture is made 'efficient' by using the economic market as a 'guide,' using science and technology to 'optimize' elements of the productive forces (farmer and agricultural improvement) and adjusting structures 'sufficiently' to utilize social and natural resources within the context of China's current social, economic, resource, and political condition. Previous agriculture policy in China is acknowledged in the articles as having been beneficial, but it is also considered to have only addressed "external conditions."[11] Efficient agriculture is intended to "form a good circulation itself which has a stronger internal developing drive."[12] Inefficient agriculture is inhibits the development of 'good circulation' and a 'stronger internal developing drive' because it is a 'deleting type' that 'over

---

9. Ibid., 575.

10. Ibid.

11. Ibid.

12. Ibid.

manages' and is 'exploitative,' all of which are word pictures of perceived externally imposed blockages. The value of having a guide, of using science as a tuning device to optimize existing conditions is the way to overcome these blockages. That is, to guide and adjust structures, in this case agricultural policy, for the sufficient utilization of resources is preferred to overly managed operations and exploitation of the available human and natural resources. The result of these heavy-handed measures is the 'deleting' or diminishing of social and economic returns.

Although the first set of relations, China/US as relations among states, is specifically focused on following a model to achieve the goal of becoming a developed country, the contrast between efficiency/inefficiency emphasizes the need to improve internal skills or advantages rather than excessive reliance on outside management or an outside model. The concern by the architects of CEA comes from a recognition of the hazards of blindly following all aspects of models, of being overly influenced by external factors. Over reliance on external control or influence is inefficient because it is a failure to recognize internal or local conditions and variations. External conditions can guide, and even optimize, but they cannot replace, nor solely sustain that which is internal and unique. As Cheng Xu put it, China 'unwittingly' followed the Western model and is now left with inefficient agricultural production because local and uniquely Chinese conditions and abilities were overlooked.[13]

The last set of relationships is between the market and the environment. According to an article by Liu Sihua, a market economy is an economy of the "deployment of resources" that are deployed by "independent enterprises" by means of a "market system" based on "law" and "equal competition."[14] Liu goes

---

13. Cheng Xu, "'Hard' and 'Soft' Restraints for China to Sustain Agricultural Development and to Follow the Conventional Modernization Approach of Agriculture and Deserved Alternative Way," 408.

14. Liu Sihua, "An Approach to the Problems Concerning Environment Protection and Ecology Construction in Socialist Market Economy," *Ecological Economic (Sheng Tai Jing Ji)*, 3.

on to point out that environmental protection is limited in this system by the competitive need for inexpensive resources. The consequence of this process is that "social production and social life have exhausted the natural resources and the environmental quality, which results in the negative value of the eco-system and also means the degradation or loss of the usability of the ecological environment."[15] Environmental protection is a holistic endeavor that is manifested in the "macrocosm," that is "long-term in time" and "spatially unalterable."[16] Liu says that people can no longer afford to think of natural resources as somehow separate from their environment, and being products of nature resource renewal is not quick, being limited and determined by geographical areas and conditions. The market economy is an economy of currency in which environmental protection is "temporarily put aside" in favor of trying to "maximize benefits."[17] In contrast to the maximizing function of the market, "the foremost output of the environmental protection is the social benefits" of having adequate resources. This however requires a relatively long time and is difficult to do given the principles of the market economy outlined above.

The relationship is one of contradictions, primarily the contradiction between the individual, short-term, usury nature of the market economy and the need for long-term, slower growth, that results in ecological protection and renewal leading to overall social benefits. There is in this set of oppositions the recognition that the market economy does have social benefits and is connected

---

15. Ibid. This tension is also being explored in Western literature, for example Matthew Alan Cahn, *Environmental Deceptions: The Tension Between Liberalism and Environmental Policy Making in the United States*, (Albany,NY:SUNY, 1995); Andrew McLaughlin, *Regarding Nature: Industrialism and Deep Ecology*, (Albany,NY: SUNY, 1995); and Paul Thiers, "Developing a Market for Sustainable Agriculture," paper presented at the 1995 World Sustainable Agriculture Association conference in Beijing, 1994.

16. Ibid.

17. Ibid., 4. This can also be seen in our own American legislative battles where environmental protection is pitted as an inhibitor of economic efficiency and growth.

with the environment. The natural resource base is "indispensable for the growth of the economy because it can provide the necessary cheap and excellent resources for economic development."[18]   However, "dependence on petrochemical energy, will damage the ecological balance itself, which will result in the deterioration of the environment."[19]   Damage occurs through the excessive production and application of chemical fertilizers and pesticides and is seen in the increased pollution of the air, soil, water, and food that is consumed. The focus on the relationship between the market and the environment is symbolic of China's current marketization policies. It addresses the question of the ideal relationship between the individual and society - how does one balance the use and distribution of limited resources to support individual short-term needs with long-term social needs?

What do these relationships and their presentation by the authors of ecological agriculture say about the conflicting value sets that must be rectified? The contradictions that need to be resolved are those between developed, advantaged countries and developing, disadvantaged countries, between inefficient short-term, deleting growth and efficient, long-term growth, and between short-term individual monetary gain and long-term balanced, sustainable social gain. ·The underlying question of ecological agriculture is one of how to establish the most 'efficient' form of development that will resolve the contradictions that these relationships establish.   Following the model of development established by the United States and Europe does not work. Individuals, whatever the unit of analysis as individuals or nations need to develop their inherited abilities and use the chosen model as a guide to follow its essence or ideal, rather than to follow blindly all aspects of it including the

---

18. Yang Wenjing, 12.

19. Ibid.

historical path that lead to the emergence of a given model. This comes out more clearly as one examines the language used to describe the contradictions.

The structure of the articles is in a narration that utilizes a scientific discourse to present a scientific consideration of social and political issues by using the language of the computer age or with the logic of Western economics. The concept of 'hard' as in 'hardware' is used to describe external factors some of which are unchangeable or 'constants' that remain unchanged. For example, limited energy resources, especially petroleum, limited water supplies, increasing pollution, government policies (rural industry, opening up the markets, pricing policy, lightening the peasant's burden, etc.), following the United States as a model, reliance on the governance of a market system are all external, 'hard' factors that are influencing the path of development.

The Chinese word for hard, *jian* also can mean resolute, solid, or firm and refer to attitudes as well as physical properties. The sense of difficulty conveyed by the word *jian* is reenforced by the adjective used to describe the constants. Here are some examples to illustrate the point: intensive use of petro-chemical agriculture creates a 'gloomy perspective,' 'the serious shortage of water ... is a bottleneck of input,' 'serious pollution,' 'severe shortages,' 'deleting agriculture,' 'inefficient agriculture,' 'activities governed by the market are contradictory to the environment.' Yet although there is a distinctively negative connotation to external factors, they are not isolated conditions to deal with, they are not themselves seen as a single problem separate from other factors. These conditions represent environmental, political, and resource issues that cannot be quickly if at all dealt with to solve each issue. If this were the case, the problems would be easier to deal with because they would appear to be more easily defined and therefore resolvable. In addition to these 'hard' constraints, there are 'soft,' internal constraints to development that must be addressed in conjunction with the 'hard' problems. The external problems are one part of a set of changeable and unchangeable contradictions.[20]

As in a computer system, good hardware must be combined with appropriate software to have a complete and operational system. Soft, or *ruan*, means flexible, supple, or pliable and as in the computer analogy it is equated with what is internal and vital for the proper functioning of the hardware. Soft is 'internal,' 'invisible;' it is 'philosophical attitudes,' 'technical innovation,' 'institutionalized practices,' 'supporting condition,' 'efficient,' 'renewed ideas,' 'long term,' 'environmental management' and 'must adjust in correspondence to the external or observable changes.' Soft is the human component of development; it is the localized, domestic and unique character of localities, regions and countries (developed countries as external/hard and developing countries as local/soft or changing). The soft, or internal, does not require external 'governance' or 'extensive management' instead, once the appropriate changes occur, it operates with 'guidance,' 'optimizing' influences, and 'sufficient' or adequate use of resources. It is by improving the software that the more rigid structure of the hardware can be overcome and worked with to one's advantage.

According to the proponents of ecological agriculture, the soft components, the software of development, have taken a backseat or been ignored at the expense of developing the hardware. In this condition the soft component's underdeveloped state is a part of the problem with continuing on the current path of advancement. So they are not at this time 'good' as in contrast to the negative images of the hard, but they are flexible and pliable, can be changed, and need to be changed into the positive force necessary to beneficially interact with the resolute and firm components. The 'internal drive mechanism' needs to be 'self-developing,' and this is a long-term, slower growth investment that has greater

---

20. Mao's "On Contradiction" continues to be useful for understanding the dialectical nature of relationships, and as I was told by a Chinese exchange student in the United States is one of two still influential works of Chairman Mao. The other is "On Practice."

overall social dividends than the current short-term, rapid growth policies and models being followed. What this means is that China's human resources, as sources of labor, innovation, and public policy, need to be brought to bear to deal creatively with the specific political, environmental, and resources constraints that China now finds itself in. These are conditions that are historically unique to China and that China must resolve in its own way, learning from the model of others, but not relying on others who have historically different experiences. These authors believe that China's leaders have become too enamored by the American model of development and are ignoring China's own internal potential at the expense of sustainable social and economic development. It is also a call for less heavy-handed management of local conditions and resources by policy makers too far removed from those conditions.

The second set of 'soft' values, the internal mechanism, is the one that can be improved upon to achieve the desired transformation of the current path of development. The belief and hope are that by focusing on China's internal resources and conditions at a local level change can occur. This will be a transformation from within, from within existing conditions, from within existing knowledge and resource bases, and from within historical experience. Most importantly the transformation will result in a 'scientific' - implying rational and calculated - plan for development. By emphasizing its scientific nature, the desired goal of 'efficient' agriculture will be based on the same premise that made implementing the current 'hardware' so attractive; it is based on science and technology, on the model and ideal of being advanced, developed, and rational, in a word - modern. The interest in 'hard' science is its seeming ability to be more objective, less emotional than the 'soft' sciences of society, politics, and culture.[21] The use of scientific discourse is an affirmation that the 'soft' human aspects of

---

21. See M.E. Hawkesworth, *Theoretical Issues in Policy Analysis*, (Albany :SUNY, 1988) for a discussion of rationality and irrationality as positive and negative attributes that are a product of the fact/value dichotomy established by positivist definitions of science.

development can be rational and therefore modern. What makes the ideas scientific in practice are the sophisticated linkages, the 'complicated systematic engineering,' between natural and human resources, between government and society, and the relationship of science and technology to all of these. It is this "complicated systematic engineering" that the ecological agriculture researchers are working so hard to explicate as they provide the scientific basis for traditional methods, to explain in the modern terms of agronomy, ecology, biology, and economics why they work and how they can be expanded and improved.

Having illustrated the sets of oppositions presented in the literature, I will now consider who is described as the agents of change that will be doing the transforming, how will it be done, and what will be the result of this 'complicated engineering?' First, "It is the peasants who are the key to realize the objective ... so peasant's scientific and cultural quality directly affects its developing fate."[22] Secondly, it is the model of ecological agriculture "that was originated by some scientists and rural technicians [which] can be regarded as a farsighted searching for sustainable agriculture and the alternative approach for conventional modernization."[23] Lastly, it is the

> government, being the representative of the whole nation, [that] can solve all the contradictions in the deployment of resources [contradictions between individual gain and social benefit] so as to seek the coordinated development of economy, society, and ecology.[24]

The point of emphasis for initiating change is on the role and improvement of the individual, 'guided' by a model, local market conditions, and within the coordinating, macroperspective of the government. Only motivated, educated individuals will be able to put the benefits of the scientific and integrated model

---

22. Wang Zhengping, 578.

23. Cheng Xu, 415.

24. Liu Sihua, 9.

164

of ecological agriculture to work on their own farms or in their agriculturally related businesses and to improve the 'software' and adapt to rather than rely on the 'hardware.' It is the individual who knows and understands local needs, abilities, and conditions. It is the individual farmer who must understand the new techniques and be willing to apply them to the planting of crops and the raising of animals. As illustrated through the implementation of ecological agriculture and its supporting literature to motivate and educate individuals, it is necessary to have a sound model and supportive policies that maintain individual gain, and also connect the individual to the national process of social improvement.

I believe that by shifting the unit of analysis from a person to the state, it is the individual state, rather than the individual farmer, that must be able to adapt domestic advantage to international constraints. The following quote by a group of Chinese researchers makes this clear.

> Sustainable agriculture has come to [the] stage of transforming the concept into concrete work. The formed strategies are being combined with practice by the producer. The China/Canada joint project provide[s] us an opportunity of absorbing advanced experiences from [the] international world including Canada in Sustainable agriculture, enable us to combine our specific situation to form our own strategies, to promote agricultural modernization in China [*sic*].[25]

This reiterates and summarizes the role and place of a model as a form of ritual behavior, behavior that entails specific notions about the concept of the individual as being defined by didactic social relations. By working and learning, 'absorbing' from someone with 'advanced experiences,' then the Chinese can combine that which is absorbed with a 'specific situation' to form their 'own strategies' 'combined with practice' to promote 'modernization.' This is not about the direct application of what is learned. It is about looking to those who are

---

25. Lan Jusheng, Jia Rujiang and Wei Jiankun, "The Concept and Strategies of Establishing Sustainable Agriculture in North China Lowland Plain," Hebei Academy of Agricultural and Forestry Sciences, Shijiazhuang, China. Presented at the World Sustainable Agriculture Association Conference, Beijing, 1994.

where one wants to be, understanding clearly where one is now relative to one's own resources, abilities, and goal, and then learning from the experiences as well as the results of those one is following as a model.

> This is mirrored in the following statement about ecological agriculture. Ecological Agriculture reflects the marriage of Oriental wisdom and modern science and technology. Its philosophy could be summarized as: a. to insist [on] the supplement between the essence of traditional China's agriculture, particularly the principle of agricultural product's recycling and multi-trophic level utilization, and modern technologies rather than replace the former with the latter. b. to embody by [all] possible means the ancient Chinese philosophy of `bring one's strong points into full play while avoiding to be [*sic*] depend[ent] upon one's weak points.` c. to substitute talent input for physical input.[26]

A modern, 'efficient' agriculture that has become a new 'target' or a new goal for 'peasants to develop agriculture' is being achieved by following models such as that of ecological agriculture.[27] The model is scientific and beneficial because it "incorporates among space, time, and industry, so it can make full use of land, material and laborer and get more with less input, which arouses the leaders at all levels to take it seriously and greatly attracts peasant's attention [*sic*]."[28]

As the model is further perfected and disseminated and as the farmers become more educated about the model, there is a need to ensure 'efficient' progress in protecting resources and equalizing benefits. To do this requires government control and coordination. The government allows the 'market system' to operate nationally, but monitors consumption and distribution of resources to create policies that will force people to consider the costs of resource consumption. For example, according to Liu, the government should force

---

26. Cheng Xu, 415.

27. Wang Zhengping, 575.

28. Ibid., 576.

financial compensation for the costs of renewing damaged resources that is in line with the money earned from those resources. Environmental resources should be taxed, and resources should be priced so that their true cost is reflected in the cost of the product and the formation of the price. The state should also control the investment structure "to assure the proportional and coordinated development of environmental investment and the investment in production and construction."[29]

The transformation desired is a 'strategic shift.' This shift as described by Lu requires the Chinese government

> not only to adapt to the requirements of [the] market economy, to resolve the contradiction between supply and demand, to solve the problem that agricultural relative benefit [for the farmer's] is on the low side and improve economic efficiency, by increasing the employment for rural work force [to] raise peasants income, but is also the only way to realize agricultural modernization in China [*sic*].[30]

'Agricultural modernization' means resolving these contradictions and problems, establishing a goal that is different from the one now being followed, and finding new or different models to follow to achieve that goal.

To summarize, agricultural modernization is bringing the 'soft,' or human components into full play to achieve a level of development that, like the model being followed, is 'advanced.' By focusing on Chinese abilities and resources, the Chinese are achieving development that is also uniquely Chinese and because of its success will become a model for others wanting an alternative model for development to emulate. It will be a modern method of development because it will be scientifically based, self-sufficient, and economically viable, just as China envisions itself being in the twenty-first century.

---

29. Liu Sihua, 12.

30. Lu 1993, "The Way to Realize Agricultural Modernization in China is to develop the high yield, good quality, and high Efficient Agriculture," *Science and Technology Daily* 2:15 cited in Wang Zhengping, 575.

Individuals, the scientific community as innovators working to constantly improve the model, and the state are the three agents of change that must work together to perfect and implement the model. They must have a clear understanding of the inefficiencies of the 'hard' constraints, which have largely been the result of 'unwittingly' applying the model of advanced countries without careful consideration of the unique characteristics and conditions of the individual (as person, as model, as state) from a local, regional, and national perspective. Because of the required interaction and cooperation of these three change agents CEA is a comment on a vision of the future that involves the whole of Chinese society, not just farmers as an independent population segment. Further it is a comment about China as an individual country in relationship to 'advanced' countries, of which it wants to be one. My analysis of this project has found it to be a symbol of China's contemporary development dilemmas, which includes production, resource, and political identity issues that have political and cultural resonance. It is from this symbolic basis that one can conceptualize the themes entailed in the ritual of following models apart from the project and see them as components of a Chinese concept of self that affects our own interpretations.

Consolidating the Cultural Themes

What are the main cultural themes that emerge from this analysis and my analysis in Chapter 4? There are three main themes that emerge here: (1) the importance of the individual, (2) the relationship between the individual and the government, and (3) the concept of model-following as an embodiment of the first two themes that helps one to understand one perspective on the nature of social relationships and power.

First is the importance of the individual in being able to improve and induce change. This focus on the individual farmer and encouragement for the individual to take action is a part of the concept of ecological, that is, a recognition of the interrelatedness of people and their environment and the nature of relations among people. Specifically, that individuals are defined, or understand themselves in terms of the relationships that they are a part of. This is at the core of the philosophy of ecological agriculture, and it became apparent in the method of implementation, that following models is dependent upon the individual seeing the good inherent in the model, which only then leads to acceptance and implementation. The relationship between people and the environment underlies the specific techniques and methods of ecological agriculture. The relations among people are to balance the needs of the individual with the needs of the larger social group. As the philosophical foundation for a revamping of the structure of agriculture in China ecological agriculture acknowledges the need for macrolevel planning and organization, which sets up a framework for state-society relations. It is a comment on the need for a governmental body with a macrosocial perspective to guide and direct individual action, and an acknowledgement that individual support and action are necessary to achieve government goals.

The realization of the need for a macrolevel of coordination to maintain and ensure an egalitarian use and distribution of resources is a recognition by the authors of the policy of ecological agriculture that individual perspectives are limited with regard to general social well-being. Policy coordination by a government body that allows for individual flexibility and initiative yet compensates for the weaknesses and usury nature of reliance on the market is fundamental. The macrolevel view of national government leaders who are responsible for this coordination will be based partly on domestic issues and partly on their sense of who they are in relationship to other countries. Both views combine to formulate desirable goals to be achieved as a nation.

Goals set by the coordinating governmental body are achieved by establishing and perfecting models for the people to follow. This allows for individual acceptance, flexibility based on local conditions, and problems defined and solved based on individual ability and resources. As seen in the analysis of ecological agriculture as a myth, power can be maintained by creating and/or being a model. The value of the model as perceived by individuals leads to either support or lack of support. Establishing models that evoke support can turn attention away from other issues. This can be tolerated by people if following the model does result in benefits and improvement, which due to the nature of the model following process can take time, and the belief that the government is aware of and addressing the larger issues.

National government has national goals to achieve, so it uses model-following as a means of generating individual support and action to achieve those goals. Additionally, model-following is not only a means to achieve an end, but the goal is to surpass and become a model. Based on statements in interviews and in the analysis of written materials, it is clear that America, viewed as a model and China as a nation-state trying to establish an independent and alternative path of development, will itself be a model for other countries. The national government is following models set by other states that it perceives itself to be in significant relationships with. This is an extension of the concept of ecology that individuals are defined by the nature of the social relations they are involved in. This is not a smooth process. Following models, as in any implementation endeavor, generates problems to be overcome.

These are problems not only of implementing reforms but of maintaining social stability in a population whose expectations are changing rapidly and who are not presently confident in the existing government. Without social stability and a unity of purpose it is feared that chaos will ensue and the drive for modernization will stall. In the case of agricultural development the problems are two- fold: a general decline in the quality and availability of natural resources and a growing dissatisfaction among farmers with their inclusion, relative to urban areas, in the benefits of development. Ecological agriculture as a model of 'modern' agriculture is intended to address these issues, and due to the slow process of adaptation it has perhaps given the national government a little more time to follow existing monetary policies to achieve its industrial and urbanization goals. It is also a model of positive action that can be used to mollify international critics of China's environmental policy.

The Concept of Model-following

Model-following, a term that I coined myself due to a lack of better terminology, is the concept of overarching importance that emerges from this analysis. It is a ritual behavior, so it is the physical enactment of beliefs and of policy. It is a tool for socializing the individual by establishing a socially acceptable ideal to be achieved, and it reenforces the social structure through individual inclusion in the decision-making process necessary for adapting the model to local conditions. Model-following establishes a definition of good leadership, as one able to be and establish good models to follow. It defines a form of cooperation based upon paternalistic power and didactic social relations, that through education as a means of teaching the virtues of a given model, the individual will see for themselves through interaction the means for achieving the goals established by the model. Ideally this process leads to equaling and surpassing the model, which forces a reciprocal pressure on the leader to continually update and improve the qualities of the model. Using models does not require a high degree of social control; there must be a tolerance for individual experimentation with models and the time it takes for the evaluation and adaptation of all or parts of a model to local needs, conditions, and abilities.

The use of models has cultural appeal because of its historical roots in Confucianism. It allows for the virtuous leader, as embodying a standard of right that has contemporary significance for a population, to chose goals and designate models. It acknowledges the importance of the individual for implementing the model, and the identity of the individual is based upon one's social relations and the imperative for people themselves to become virtuous, to conform to the current standard of 'right' as defined by the leader and embodied in the models.

Mao recognized the importance of the individuals and of gaining their acceptance of his vision of the communist model that he was trying to implement. The following quote illustrates this recognition as well as representing the nature of the relationship between government/party leaders and the population.

> These cadres, the peasants, the jailer, the merchant and the revenue clerk were all my esteemed teachers, and as their pupil I had to be respectful and diligent and comradely in my attitude; otherwise they would have paid no attention to me, and, though they knew, would not have spoken or, if they spoke, would not have told all they knew....It has to be understood that the masses are the real heroes, while we ourselves are often childish and ignorant, and without this understanding it is impossible to acquire even the most rudimentary knowledge.[31]

Or again in "Get Organized," Mao says,

> The masses have great creative power. ... We should go to the masses and learn from them, synthesize their experience into better, articulated principles and methods, then do propaganda [teaching] among the masses, and call upon them to put these principles and methods into practice so as to solve their problems and help them achieve liberation and happiness.[32]

From the perspective of the model-following concept, Mao is realizing that the model of communism must be based upon individual and local experience; it must reflect local conditions so that the population at large will see the goodness of the model, accept its principles, and be willing to implement the tenants of the model for their own benefit and the benefit of society as a whole. In a word it is praxis. Mao's model of communism as embodying a definition of social good worked because he was able to generate local, individual support and enthusiasm.

---

31. Mao Tsetung, "Preface to Rural Surveys," in *Selected Readings from the Works of Mao TseTung* (Beijing: Foreign Language Press, 1971), 195-196.

32. Mao TseTung, "Get Organized," in *Selected Readings*, 301.

172

In the contemporary ecological agriculture literature the use of the word 'individual' is also a marked change from the equalization of all people as the 'masses' under CCP rule, and a recognition of the need for greater local variation in adapting new models. This emphasis may partly be a result of Western influences about ideas of the individual, but it also significantly recalls the ideas of the Chinese philosopher Mencius.

> Mencius, who lived from 372 to 289 B.C.,...believed that respecting, protecting, educating, helping, and enriching people would be the best way to develop the country. If people were allowed to pursue their own goals, a stable, peaceful, and developing society would follow.[33]

In line with this Mencius also believed that government is created for the people, not the other way around, and the government's inability to care for the people is grounds for change. The point is significant for two reasons, the first is that while the West may want to claim credit for introducing the concept of individualism into China we can not, and second the philosophy of ecological agriculture is one of reinvigorating internal ideas *guided*, but not driven, by external ideas. While ecological agriculture as a philosophy re-invigorates the role of the individual, it is also a marked shift from traditional Confucianist ideas

---

33. Zhiling Lin, "Traditional, Modernist, and 'Party' Culture in Contemporary Chinese Society," in Zhiling Lin and Thomas W. Robinson, eds., *The Chinese and Their Future: Beijing, Taipei, and Hong Kong* (Washington,D.C.: The AEI Press, 1994), 142. Regarding traditional Confucianist notions of individual Wing-Tsit Chan makes the following point: "Since both Taoism and Buddhism deny the self and Confucianism teaches obedience, it seems that Chinese thought attaches little importance to the individual as such. Such observations are superficial. While in traditional China the Chinese has had no personal choice in marriage, he has enjoyed absolute freedom in the choice of religion. There is a general regard for privacy, in religion as in other spheres of life. While property has been held in common, each son has had his inalienable right to inheritance. There was no individual vote guaranteed by a constitution, and yet in village meetings every male adult was a voting member by natural right. In the thirteen-century-long tradition of civil service examinations, the basis for the selection of government officials was individual merit rather than race, creed, economic status, sex, or age." Wing-Tsit Chan, "Chinese Theory and Practice," in Charles A. Moore, ed., *The Chinese Mind: Essentials of Chinese Philosophy and Culture* (Honolulu: University of Hawaii Press, 1967), 24.

about man's relationship with nature. Cheng Xu points out that in addition to positive philosophical viewpoints in support of traditional sustainable agriculture

> there also existed some negative aspects, such as bias of land reclamation, the viewpoint of considering (that) mankind will definitely 'triumph over' nature and the self-intoxicated complacency feeling for so-called 'vast territory and ample reserves.[34]

The emphasis for policy formation and implementation is on creating a virtuous model, where virtue is defined as what is 'good' within specific historical and policy contexts. A model that supports a standard of right belief and action, and paternalistic power is the basis of power that facilitates dissemination and acceptance of the model. Paternalistic power is based upon education, the giving of knowledge from those who have to those who do not, and forms the structure of power within China. This means setting goals and establishing a path to follow towards their achievement that is feasible and popularly supported. The population will tolerate a strong central government as long as it maintains this course. Conflict occurs when national goals are being achieved at the expense or one of more domestic groups to the degree that social instability becomes a threat and a real possibility.

On the part of the leader, whether a national or international authority, power is expressed through the formation of goals and rules/policy for achieving those goals. Power for those following a model lies in their interpretation of the rules and their resulting action. The rules can be followed, openly disobeyed, or passively dealt with as a means of protesting and as a means of testing to see what can be gotten away with. As described in Chapter 3, some rules, such as instituting family farming, are vague and left to local interpretation as a means of implementation. In following the rules of a model individuals, whether defined as individuals or states, can exert power in several ways: (1) they can manipulate

---

34. Cheng Xu, 413.

174

policy either through a deliberate misconstruing of the policy or by exploiting weaknesses in the new rules; (2) they breaking the rules as a means of experimenting and testing new policies; (3) they can achieve necessary production by using policy alternative banned by the leader; and (4) they are aware of the effectiveness of passive means such as sabotage or deliberate lethargy. On the other hand the governing authority does maintain ultimate power over the population at large, because there is a general deferment to the larger authority of the group. The effectiveness of peasant exertions of power is not an overall change in the policy itself but can be a qualitative change in the degree to which the policy is effective or encompassing.

What these cultural themes and cues tell us as observers of China about Chinese politics is a sense of local and national identity that is culturally comfortable and works well enough to be used to implement a nation wide policy for agricultural change. It tells us that goals and models can be potentially stop gap measures to divert attention away from conflicting policy actions. The same goals and models can also be realistic solutions to existing problems. This can not tell us if a civil-society is emerging in China, or of the imminence of democratization or marketization as American scholars are so intent on trying to do.[35] My enunciation of these themes makes no pretense of prediction but is an attempt to clarify an important component of Chinese social and cultural identity which has an impact on how the Chinese view themselves and the world around them. As social scientists we should be trying to increase our understanding of how others see themselves in the world. This is not a search for causal

---

35. The September 1994 issue of *The China Quarterly* is devoted to an overview of the state of comparative political studies of China. It is a search for clues that can lead us to predict what the outcome of China's current struggles might be, will it democratize, become retrenched in Communism by remaining stagnate or erupt into revolutionary change. In fact these are the general conclusions/ predictions arrived at by the various authors. (p.702) These are also the types of conclusions arrived at by the contributors to Samuel Kim and Lowell Dittmer,eds., *China's Search for National Identity*, (Ithaca:Cornell University Press, 1994). These conclusions do not explain and do not increase understanding; what they do say is that based on our American use and definition of the analytical concepts employed to study politics, we can't understand what we observe happening in China.

explanation, a backwards glance at why, but a search for what do policy actions mean for the other person now and how that relates to what we observe. In this way understanding can be increased.

What does this study mean for our analysis of Chinese politics? One must question the definitions one gives to these analytical concepts, to the concepts used to define the conclusions of observed behavior. Do we know the fate of the CCP? Can we predict democracy or another revolution? The answer must be no, and even if we could what good would it be if we cannot see what meanings these changes may or may not have for the Chinese. If one can consider the ecological program as symbolic of a Chinese attempt to integrate Western and Chinese ideas on a larger scale, beyond the specifics of agriculture, then the cultural themes I have outlined are also symbolic of a desirable concept of self. It is a self that ideally behaves in such a way as to maintain harmony and balance between the individual and nature and between the individual and society. These themes do not dictate a form of government but represent ideals which must be a part of government because they allow for modernization and technological growth. This ideal of self gives meaning that affects the interpretation of one's place in the world and of the experiences encountered. It is an understanding of the nature of social relationships, the power holders in each relationship and the actions authorized by the different interpretations, locally, nationally, and internationally.

The questions that should be answered are: how do these themes serve to offer a different explanation of Chinese behavior? or how our analytical concepts of the individual, cooperation, democracy changed when we consider a possible Chinese definition of them? China's relationship with the United States is as with a model and models are meant to be surpassed, the Chinese government is struggling to implement the goals of the 1973 reforms and to emerge as a new model of government with new goals for the next century. If one asks how Chinese political behavior may be explained if the Chinese see themselves in the role of being and following models and the implications that has for ideals of leadership, policy formation, and implementation, their analysis and conclusions will be different.

Using models requires a tolerance for unsettledness because adaptation or implementation problems cannot be predicted or projected. As described in Chapters 2 and 4 problems emerge and must be solved on a daily basis. What may appear to us as conflict between what is said by the Chinese and what is done, which gives way to perceived irony and distrust on our part, may not be so ironic if one considers meanings from their view. If the Chinese are trying to establish themselves as a model for other countries to follow, then they are not going to be a leader that actively leads in the name of a group as Americans would actively lead. Through their own behavior and continual self-improvement based on their standards of 'right,' the Chinese are striving to be a model that others will want to emulate. Focusing on perfecting the model is the leader's responsibility, learning from the model is the follower's.

Comparing Chinese and Western Cultural Themes

There are a number of themes that are central or make up the core of American studies of Chinese domestic politics, and as questions and observations they reflect our interests or motivations for understanding Chinese politics, for example: (1) what is the impact of Western liberalism, (2) the apparent unity of China's past culture is now becoming one of disunity, what does this indicate (3) China is too culturally diverse to speak of a single culture or identity, where do we look to see 'China' (4) is China still a Third World state and leader, (5) what can we learn about Chinese political goals by studying Chinese foreign policy, (6) China has followed the rules of the U.N. and the World Bank and wants to join the World Trade Organization, has it been and will it continue to be a cooperative team player? To answer these question American analysts look at different components of Chinese society such as elites, intellectuals, peasants, minorities, industry, agriculture, the party, history, domestic policy, international policy, technology, economics, political culture, etc. These questions ask about our relationship as Americans with China, what has our impact been, and is China still the threat to capitalism and our world vision that we once believed it was as a strong communist state. There is an underlying assumption that the Chinese will come to see the world as we see it, to embrace our values as we embrace them. Yet, what they fail to ask is what meaning do the Chinese give to our world view and to the values we embrace.

How does our understanding and analysis change if we consider the cultural themes that I have outlined in this study. I will consider the first three questions about our observations of Chinese domestic politics as one category, and the last three as a second category about our observations about China's international behavior. This discussion will bring me back to articles compiled in the book *China's Quest for National Identity* and *China's Participation in the IMF, the World Bank and GATT*. Two lines of questioning emerge from this discussion, one is how the comments and analysis of the various authors change when discussed in terms of model following and formation as a part of Chinese culture. The second is comparative, asking to what degree do their analyses reflect issues that are problematic for American culture and politics. Both sets of questions and observations should encourage us to become more self-reflective, less quick to establish dichotomies of good/bad, us/them, right/wrong. Hopefully we can become more conscious of the questions that we ask, how we ask them and of what we expect to find. What is the impact of Western liberalism on modern China? This is perhaps the most significant question that we ask about China or that the Chinese, as seen in the ecological agriculture philosophy, ask themselves. Michael Ng-Quinn believes that Western liberalism is the largest challenge to traditional Chinese identity.[36] Traditional Chinese identity as he defines it is one based upon unity and culture, it "offered stability and helped sustain the Chinese state regardless of the circumstances by unifying the

---

36. Michael Ng-Quinn, "National Identity and Premodern China," in Dittmer and Kim, 32-61.

population, at least psychologically."[37]   Given that ideas about federalism, Taiwan's independence and the possibility of multiple identities as preferable alternatives to an "obsolescent, singular national identity," China faces two challenges from its exposure to the West.[38]

First is the contradiction between the need for the "acceptance of certain degrees of hierarchy and authority" to maintain the unity of a nation as large as China and the 'fact' that the prevailing ideology outside of China "where the largest number of visiting Chinese students and professionals can be found" is Western liberalism.   The values of liberalism (freedom, individualism, and pluralism) are "precisely the opposite" of the values of required of unity (acceptance of hierarchy and authority).[39] The second challenge is based on Ng-Quinn's observation that "culture helps make a state distinctive, whereas technology and a capitalist world economy have had a homogenizing effect across state boundaries."[40]   The result is a dichotomy of values that is presented as an either/or choice. "Materialism and efficacy have subjected dignity and pride to critical reevaluation and redefinition. Given these two challenges, it remains an open question whether a Chinese national identity based on unity and culture will be able to move forward in its long journey toward 'maturity.'[41]

The question of Chinese social and political unity is of central concern in a variety of forms. Lynn White and Li Cheng examine the regional, national and global identities they found in Guangdong, Taiwan, and Hong Kong.[42] The identities of these coastal groups poses a challenge by "threatening to modernize

---

37. Ibid., 57.

38. Ibid., 60.

39. Ibid.

40. Ibid., 61.

41. Ibid.

42. Lynn White and Li Cheng, "China Coast Identities," in Dittmer and Kim, 191.

the old ideal claims to legitimacy which Chinese rulers and intellectuals have long held to be the moral basis of their polity."[43]   Because these communities are coastal they are the points of interface between China and the West and it is for this reason that they are seen as a challenge to the old ideals. "The periphery is in fact more democratic than the center, because a greater variety of people - not just officials, technocrats, and intellectuals - have more wealth, prestige, and power there."[44] The people in these more democratic regions see the Chinese state as the "guardians of a moral order rather than the outcome of a process."[45]

The increase in democracy, per their materialistic definition of it, is a threat to the national leaders owing to the traditional role of the peripheries to protect and strengthen China while absorbing new notions and resources. This is the general conclusion of the multiple identities argument, the realization that the centralized, authoritarian Communist regime is poorly equipped to deal with such diversity and that pluralism as democracy is the most viable solution. It is interesting to note here that democracy is equated with the results of marketization, which is not based upon a democratic process, but it is expected that the government should be. And we must ask that even if the government were the result of a process, implying a democratic process, wouldn't it be charged with protecting the new rather than the old moral order?   There is an equation of democracy with markets, of multiple identities with pluralism, of the new identities at the expense of and separate from the old. The 'old' CCP order is only 45 years old, a blip on the screen of Chinese history, it is not far fetched to ask what exactly is 'old,' and what parts of the old are being maintained even if in a modified form? And what is the role of government if it is not to protect social order, which is inherently a moral order of some form?

---

43. Ibid., 191.

44. Ibid.

45. Tu Wei-ming, "Cultural China: The Periphery as the Center," *Daedalus* 120 (Spring 1991), 15-16 in White and Li, "China Coast Identities," 192.

Instead of using the concept of emerging liberalism in China, we can see the current domestic situation as a period of redefining relationships within and without China. It is a time of formulating a new model, and as such is a period that fosters and tolerates change and uncertainty. Should the post-Deng leadership be able to articulate the new order of relationships in a culturally sophisticated and politically meaningful way, the CCP can maintain viability. This means an articulation of existing contradictions in society and an offering of policy as models for overcoming the contradictions at the national, regional and local levels. It also means that the existing and any new leadership must clearly show itself as worthy of defining the models necessary to overcome the contradictions they have defined. This may or may not mean an emerging basis for Chinese style democracy, but we can not assume that pluralism and open markets are the only choices, or over estimate the external influence of Western Liberalism.[46]

Xiaoxing Han concludes his article "Democratic Transition in China" with the following, relevant quote:

> History has shown us again and again, and will show in the future, that the discontent of the majority does not necessarily mean inevitable change of a regime, and that a regime change inspired in some way by anti-dictatorial demands does not necessarily produce a meaningful democracy.[47]

What must be considered if we are to fully appreciate the changes taking place in China is the question of what would be 'meaningful' change, culturally, historically, and politically. By shedding the confining skin of our familiar

---

46. See Prasenjit Duara, *Culture, Power, and the State: Rural North China, 1900-1942*, (Stanford: Stanford University Press, 1988). Duara's research illustrates the limited effect of Western influences on change in China, and highlights how the cultural nexus of domestic factors played a much larger role.

47. Xioaxing Han, "Democratic Transition in China," in Roger V. Des Forges, Luo Ning, Wu Yen-bo, eds., *Chinese Democracy and the Crisis of 1989: Chinese and American Reflections*, (Albany,NY:SUNY, 1993), 239. See also Roger V. Des Forges,"Democracy in Chinese History," 21-52 in the same text.

analytical concepts we are free to ask new and potentially more insightful questions.

If our search for clues or hints of democracy internally have been based on looking for cracks in China's authoritarian control, what has our search for China's external or international changes revealed? One of China's most vaunted claims is that of being a Third World state and the desire to be a leader of the third world. Using the criteria of poor, non-Western, nonwhite, nonaligned, and in Asia, China is still "objectively Third World."[48] Why then is China's claim to being of the Third World as an identity suspect? Van Ness summarizes four reasons based on his observations. First is China's "national pride in being Chinese, an identification with the glories of China's past, and, one might say, a typical Chinese cultural arrogance. All of these traits point to a sense of China's uniqueness rather than of shared Third World characteristics."[49] The result is that Chinese invocation of a Third World identity, which is built upon feelings of national humiliation at the hands of the Western nations, does not necessarily create a "sense of common bond the Third World."[50]

Van Ness's second reason is that "the Chinese have had great difficulties on a person-to-person basis translating their official policy of Third World solidarity into interpersonal cooperation and harmony. Chinese racist attitudes toward African blacks studying in China are a notorious example."[51] The Chinese expect that "others will defer to PRC leadership," and in their behavior in the United Nations "China behaves not as a typical Third World country."[52] Thirdly, Van Ness notes that China's "fast track to wealth and power by cooperating with

---

48. Peter Van Ness, "China as a Third World State," in Dittmer and Kim, 212.

49. Ibid., 213.

50. Ibid.

51. Ibid.

52. Ibid.

the capitalist West" lies in contrast to earlier policies of opposition to imperialism, self-reliant development, South-South cooperation, etc. The exception is to lay claim to a Third World identity for propaganda purposes which serves "to tap basic feelings of Chinese nationalist sentiment and keep alive the option of a renewed effort to build coalitions among the countries of the Third World if necessary."[53]

Several important questions emerge here, such as was the Third World ever unified as a singular power, except in the minds of Americans and Russians as we played the Cold War game? The countries of the Third World are not homogenous, without individual histories, prides, and interests, but they are still "objectively Third World" as Van Ness defines the term. They have at times come together to form groups or blocks to voice their concerns and desires, for example OPEC, or the Group of 77 and the New International Economic Order, but they have not transcended the many boundaries that separate them to form a single political force. An international sense of separate economic, social and political identity is just as powerful in its symbolic and interpretive effect as is a more familiar political bloc of power. The second and third points are inter-related in an interesting way. If Chinese racist attitudes create difficulties for translating policies of solidarity into cooperation and harmony, would they not also have implications for Chinese attitudes towards White Europeans, the foreign devils, the very source of China's national humiliation (which is real and not a fabrication of the CCP propaganda machine)?

It would at first glance seem not so, since China is 'cooperating' with the West through participation in joint-ventures, joining the IMF, the World Bank, accepting long-term loans, supporting foreign tourism, importing foreign goods, and sending students and scholars abroad, they must identify culturally and racially more with the capitalist world. Given that this list of cooperative actions applies to most countries that could be considered as Third World countries, are

---

53. Ibid.

they all cooperating in an atmosphere of harmony and equality? If cooperation is the antonym of opposition then an entire range of behaviors that are passive forms of opposition, or momentarily advantageous interactions have been overlooked. During World War Two Americans realized the need to "hold hands with the Devil," to cooperate with Stalin in order to achieve our war time aims, and Uncle Joe temporarily became more like us rather than the ideological enemy he had been and would be again. If underlying prejudices and preferences do not change in spite of outward appearances, such as Van Ness's example of China's racism (Africans are invited to China in spite of this racist attitude), we must ask how these views affect what we now call cooperation. Lastly, we can not apply one measure of cooperation such as China's racism to one area without asking if that measure applies to other situations and for other cultures. Does American racism define or measure our levels, or degrees of international cooperation, and why does it for the Chinese?

How does the concept of model-following as the major cultural theme to emerge from ecological agriculture help us to see from a different perspective the issues of China's contact with Western liberalism which manifested in the recent American realization that domestically China is not a unified whole, or that China's identity in the international community is neither Third nor First World? One of the goals of Ecological agriculture is to preserve what has worked well in China, what is Chinese, while improving and supplementing it with the most useful aspects of Western ideas and methods. By using models an authority is established that defines and encourages the implementation of ideal goals, methods, and values. There is an acceptance of this paternalistic authority as long as what it teaches is of value to others, a value which is determined through practice and observation. Those that create the model, the leaders, recognize that it is only through the acceptance and the work of the individual and the ability of the individual to adapt the essence of the model to their own local conditions that the goals displayed through the model will be achieved. There is a great deal of latitude for choice, innovation, and experimentation on the local, individual level while still acknowledging the authority vested in the model. Further, because the model is used to achieve national goals, because large portions of the population are participating in, learning from, and following the model a sense of national community, purpose and identity is created, fostered, and maintained.

James L. Watson wonders whether or not "the fact that China no longer has an agreed upon set of rites mean that it no longer has a unified culture."[54] The dichotomy that Watson establishes is between Western and traditional Chinese approaches towards culture. In traditional China and in the contemporary rural areas Chinese culture is based on correct practice, upon li. In studying death rituals Watson summarizes that "following correct forms ensures that one is playing the game of culture by civilized rules and, in so doing, one becomes Chinese."[55] Following the correct form allows for local variations in belief and to a degree in practice while at the same time maintaining a national cultural identity. This is in contrast to an "obviously Eurocentric view of culture and national identity" that is based upon content as more important than form. "For most Westerners what is 'inside' is what counts."[56] The extension of this is that as Westerners we must come closer to sharing beliefs to maintain cultural cohesion and identity. Watson observes that the disappearance of a unified cultural tradition is expressed in the diminishing participation in funerals and weddings according that follow the same basic form, particularly noticeable in urban areas where the traditional rites are described as "'feudal' (even humorous) vestiges of a rural past."[57] From this observation comes the questions "Will a new culture, based on a new set of rites and unifying symbols, emerge from the old? Or have

---

54. James L. Watson, "Rites of Beliefs," in Dittmer and Kim, 102.

55. Ibid., 99.

56. Ibid.

57. Ibid., 102.

the Chinese people changed the ground rules of cultural construction and begun to emphasize internal beliefs and attitudes rather than external form and ritual?"[58] Given Mao's failure during the Cultural Revolution to change people's inner beliefs, Watson infers that the following form, li, is still of significance in China and that new rites and symbols will emerge from the old.

Following models is a form of li, of correct behavior. By following a model one participates in the community while maintaining and improving upon individual conditions through individual effort. Diverse identities, diverse cultures and economic abilities have been historically tolerated, as long as certain key elements are followed or implemented that serve to connect the individual to the whole, establishing and maintaining a larger collective identity of being Chinese. In trying to asses the meaning of China's contact with Western liberalism, the increasing disunity observed, (which is not a new phenomenon in the context of Chinese history) we should also be looking for those rituals and metaphors that are persisting or being created that link people together in a meaningful way.[59]

Using Van Ness's outline of Chinese foreign policy making as a pattern of policy making, we can see the model following themes in action. Chinese foreign policy making is described as a "comprehensive design, ... that begins with an analysis of the international situation. On the basis of that analysis the (foreign policy) line prescribes a strategy for dealing with the principle problems that the analysis has identified. It is a paradigm or logically integrated model of foreign relations containing prescriptions for political-strategic policies and international economic relations."[60] Emphasizing the model following ritual as presented in his description, we can see the making of policy based on the nature of China's

---

58. Ibid.

59. See Wasserstrom and Perry, eds., *Popular Protest and Political Culture in Modern China: Learning from 1989*, (Boulder: Westview Press, 1992).

60. Van Ness, 201.

relationships to others at a given point in time, and upon a concept of self as defined by those relationships. The contradictions that emerge in the analysis of relationships define the problems to be dealt with, and the policy itself prescribes policy action based on a model of appropriate action for given relations. Taking what we know of following models then there is a great degree of latitude in the actions taken, and in the beliefs maintained, as long as the spirit of the model is followed, which again is a form of li.

With respect to China's position as a Third World leader Van Ness notes that "if indeed communism is dead as a viable alternative road to modernization, the idea of a third way loses all political meaning."[61] In fact it implies that there is only one way, capitalism. To support this it is pointed out that of the three political groups within China, the conservatives, the reformers, and the radicals, the idea of Third World identity seemed "positive or constructive" only for the conservatives. "By contrast the reformers identified with the east Asian newly industrialized countries (which are the exceptions rather than the norm of Third World development) as their implicit model; and the radical aspired to China's full participation in global civilization, partaking of the best of culture wherever it was to be found rather than identifying with a more parochial and limited notion of national identity."[62]

This observation says more about the importance of self-identity in relationship to others and following models than it does about China's budding capitalism. The reformers and the radicals have been able to redefine the nature of China's international relationships in a way that the conservatives have not. Yet they are both looking for models to follow, models that are symbolic of where China wants to be in the future and a way to get there that will be uniquely Chinese, allowing China to maintain independence and dignity. Learning from

---

61. Ibid., 212.

62. Ibid., 209.

well-chosen models does not mean becoming identical to or with the model. China's "separate peace with the global status quo" does not mean that China "has shown that it wants desperately to escape from being Third World!"[63]   As was described to me in the Jiangsu Agriculture and Forestry Department, Office of Foreign Affairs and Economic Relations, (this office works with the World Bank and other international agencies setting up projects in Jiangsu Province) China is a model of interaction with the World Bank for other developing nations to learn from.   China is illustrating through action an alternative way to avoid debt and dependency and, in this way, is establishing itself as a model to be followed and therefore as a leader capable of defining and creating a separate identity within the existing order.

China can claim to be Third World as long as it does not behave in a way that is by its own definition imperialist or since 1989, American. To dismiss this identity as rhetoric without asking why it seems to recur and why it continues to be effective as a symbol of Chinese nationalism is to overlook the subtleties of Chinese identity in the world and of our inability to accept the self-definitions and perspectives of others. To examine the rhetoric and symbolism of this concept requires exploring the meanings of its definitional components more closely. For example: What is the long term vision of the reformers and of the radicals for China in relationship to other nations? How do they define the essence of the models that are now relevant for them, can they be surpassed? How was the question of Third World identity put to them when the research was being conducted? Is there a new metaphor that has the same or similar leadership connotations? Are the claims of identity and purpose any more or less rhetorical? How do they compare?   It is possible that as a third force the idea of being a Third World state has little utility, but that interaction with the West does not mean the loss of a separate leadership identity or of separate long-term goals. If as Van Ness states capitalism and technology have a homogenizing effect, is it

---

63. Ibid., 212.

reasonable to assume that the Chinese want to be homogenized? The goal of following a model is self-improvement by adapting what is essential to the model and then over time to become a model for others to follow.

By interacting with the existing international economic and political system China has acknowledged the established order of relationships, and defines itself in terms of these relationships. Power is didactic, designed or intended to teach, and as we will see in Chapter 6 the American discussions of the international system we too see it as a teaching and socializing tool, in which economic and political power are use to 'encourage' 'correct' learning. Only by maintaining a sense of identity that is based on learning from and yet maintaining a philosophically separateness can the Chinese see themselves as able to offer a continually improving leadership position to other nonindustrialized countries. Only by learning from those that have been deemed models, namely, the United States, can the Chinese improve their own capabilities and eventually surpass those that are now models.

CHAPTER 6

DOMESTICALLY INFORMED INTERNATIONAL ANALYSIS

This chapter brings me back to the original research question posed in Chapter 1: how can the study of domestic politics inform current mainstream American analyses of international behavior? How can the concept of model-following as a domestic theme of Chinese culture and a conception of power and social relations affect American understandings of Chinese cooperative behavior in international organizations? All international acts are domestically interpreted and given culturally specific meanings. By examining domestic policies as interpreted texts one can learn about cultural cues and their meanings as "symbolic forms through which each community tells itself and other audiences something about its identity as a polity."[1] Following this premise, the levels-of-analysis problem becomes a conceptual problem centered on the analyst as one of many interpreters and is not only a problem of 'correctly' describing what is observed as states interact in the international arena.

I will begin this chapter with a brief discussion of how I appropriate the domestic cultural themes that I have labeled with the concept of model-following as a useful tool for providing a different understanding of Chinese international behavior. I will show how this discussion fits into the current debate on levels of analysis and cooperation as defined by Alexander Wendt. I will enter into this debate first with an examination of the definitions of the concept of international

---

1. Dvora Yanow, "The Communication of Policy Meanings: Implementation as Interpretation and Text," in *Policy Sciences* (Spring 1994), 57.

190

system and its function as defined by Kenneth Waltz, then show how Robert Keohane's concept of cooperation emerges from Waltz's definition of system and becomes a guiding principle for Harold Jacobson and Michel Oksenberg's analysis of China's participation in various international organizations.

I then specifically focus on Samuel Kim's analysis of Chinese behavior in the United Nations.[2] I have chosen Kim's study because it is the only detailed study I have found of Chinese participation in an international organization, and it is one that also reflects the currently predominante definitions of cooperation used by international relations theorists such as Robert O. Keohane.[3] Kim's study is valuable for two reasons: (1) it provides enough detail about Chinese participation in the U.N. that the material he presents can be reconsidered with the concept of model-following as I have describe in this book, and (2) he begins the study with a detailed description of a Maoist epistemology that he believes helps to explain culturally what he observes at the UN, yet he fails to use this information to inform his use of neorealist analytical concepts, concepts which shape culturally specific interpretations and concluding observations. My reconsideration using the model-following concept is one that arrives at an alternative set of conclusions. These conclusions are different, and important because they are the result of having used an interpretive framework that is based upon the domestic Chinese concepts that through my analyses in Chapters 4 and 5, I have found to have political and cultural significance for the implementation of policy in China.

---

2. Samuel Kim, *China, the United Nations, and World Order* (Princeton: Princeton University Press, 1979).

3. Harold Jacobson and Michel Oksenberg consider China's accession into the IMF, World Bank, and GATT, but I have found Kim's analysis to be more useful for learning about how the Chinese participate as members, that is how they see themselves and interpret the rules of international cooperation. The goal for the Chinese was to become members of these institutions, so I believe that to examine how they became members tells less about a Chinese sense of international participation and cooperation than does a look at their actions as accepted members. Harold Jacobson and Michel Oksenberg, *China's Participation in the IMF,the World Bank, and GATT: Toward a Global Economic Order* (Ann Arbor: University of Michigan Press, 1990).

I will conclude this chapter and this book with a brief examination of the recent tensions during the Taiwanese elections and the types of questions that my conclusions should lead policy makers and researchers to ask. My believe is that if these concepts have import for the Chinese domestically as a reflection of an aspect of political identity that affects our analyses of Chinese domestic politics, then they may also have implications for use observers of Chinese politics to interpret China's actions internationally. In using the model-following construct, domestic politics is used for broadening international understanding by challenging the prevailing analytical concepts used for interpretation.

My point here is that if analysts are to understand the international system and processes it is necessary to ask what meaning others participating in the international arena give to their interactions. By asking this question I am confronting the micro/macro issue as a problem that needs to be reconsidered. The micro/macro gap, or the levels-of-analysis problems in international relations discourse, is an interpretive gap between American academic conceptualizations of the international system as one domestically based interpretation and other possible domestic conceptualizations.

The Appropriation of Cultural Themes

In this book I have analyzed the implementation of a Chinese domestic policy as a means of learning about Chinese political culture. My analysis rests upon two assumptions, first that language is the carrier of all cultural and historical influences, and so understanding originates in the linguistic experience of the world, and second, that belongingness to the world is the interpretive experience itself and all understanding is mediated by interpretation. The following quote from Michael Shapiro reenforces the point:

> To understand and analyze conduct, we must uncover the system
> of constitutive rules that creates different meaningful episodes and
> events, and kinds of conduct. These systems reside in discursive
> practices and these discursive practices can be the data whose

interpretation would reveal policies that allocate and legitimate various kinds of control.[4]

These systems of constitutive rules include the discursive practices of the social sciences and the definitions and models used for the categorization, labeling, and interpretation of conduct.

Hermeneutic reflection consists of the dialectic of interpreting the meaning of the research data as texts (my linguistic analysis of interviews and written material), by reflecting on the parts as themes (metaphors and rituals as indicators of culture), and moving to the meaning of the whole in relation to a respective theory (in this book it is comparativist theories of modernization and international relations theories of cooperation). Finding meaning is a process of combining description, thematic interpretation, and metaphoric insight as the product 'given' to the researcher by the data. This is the process of learning about and appropriating knowledge of and about other cultures, conceptualizing what is observed by using the language of social science and then applying what is learned to generate new questions and insights for application or research, applying what has been learned to what is observed through participation in the world.

My analysis is of the implementation of a domestic agriculture policy. This policy was analyzed through the thematic interpretation of the project title as a metaphor, the implementation process as a ritual, and the policy generally as able to support contradictory realities and policy goals. A number of cultural themes and political cues emerged which I have collected together under the concept of 'model-following' as a metaphor for a Chinese sense of social structure and power which I relate to political identity theories of comparative politics and theories of cooperation in international relations.

---

4. Michael Shapiro, *Language and Political Understanding the Politics of Discursive Practices* (New Haven: Yale University Press, 1981), 131.

Model-following is a concept that is based upon my empirical observations of the implementation and discussion of the policy of ecological agriculture. It reflects an element or a perspective of Chinese culture. I have outlined this in detail and will not go into it now; suffice it to say that the concept of model-following is historically grounded and presently active and reflects uniquely Chinese conceptions of social structure and political power. Rather than plugging what I found from my research into a preconcieved model or definitions of my own or of another's making for verification or falsification, I use this concept and its Chinese-based definitions as an analytical tool on its own merits.

This is not dissimilar to the ways currently vogue concepts from economics or other sciences are used to guide the analysis of international relations. Sociological, economic, psychological, and concepts from the natural sciences such as chaos theory are used as analogies for guiding definitions and hence the questions asked about what is observed in the realm of international action and what those observations mean. These theories are based upon empirical research conducted at an individual level of analysis about various types of human interaction. For example observations of the way people interact economically leads to generalizations about competition for resources and the power of the profit motive. Micro-economic theories of individual behaviors have been appropriated by Kenneth Waltz to explain conceptually state political action at the international level. Erik Erikson's psychological studies of development in children has been appropriated by Samuel Kim as a means of understanding and explaining Chinese political development.

My empirical observations of a Chinese political process, through the implementation of ecolgical agriculture, has not only given me an understanding of an important aspect of Chinese politics and culture, but it is now a part of my own experiential and knowledge base. I have refined through analysis what I have observed and now have a conceptual framework that is my own. As a part of my own knowledge these concepts have broadened my understanding, and I have to ask myself how this knowledge may provide insights into other areas of Chinese politics. Model-following is not an idea that I can only apply to Chinese domestic politics; rather it raises questions of meaning that I can ask of many situations involving China. When I read literature on Chinese development or on Chinese international acts I can not help but ask, "how does the concept of model-following help me to understand what I am seeing from a more Chinese perspective? What questions does it generate that may more adequately reflect a Chinese interpretation of a given political situation?" I will use the recent U.S./Chinese tensions over the Taiwanese presidential elections as an example in the conclusion of this chapter.

I see my use of the concept of model-following as a metaphor for one part of Chinese culture as different from the appropriation of economic concepts or psychological concepts in two ways: it is not based upon observations made in a familiar Western culture, and it is based upon asking questions of meaning. For these reasons I believe that my concept of model-following is different because it is an attempt to capture a Chinese based definition of politics and social identity and so does not begin from within existing social science definitions of politics and identity. This knowledge is mine however, and when I apply it to my understandings of international cooperation theories it serves to challenge those existing definitions and provide different understandings. These different understandings are important because they will have the potential to allow policy makers and analysts to make different assumptions and generate alternative discourses that I believe can facilitate cooperation and conflict resolution by stimulating new perspectives.

Why do I want to apply this knowledge to theories of international cooperation? What is at stake here? The great question of international relations theory is how to understand the relationship between domestic and international levels of analysis, what is the influence of each upon the other. The purpose in asking this question is to try to come to grips in some fashion with the processes of change and transformation in the international system, broadly defined as the relations among and between state and non-state actors. The field of international cooperation is the most fruitful for working in towards this end because it is through cooperation that interaction occurs in nonwartime, nonviolent situations. It is believed and hoped that the processes of cooperation can lead to greater political understanding and peaceful interaction among the diverse cultures of the world. What I hope to establish through this book is that understanding these processes is not only about how and why they occur but that they are based upon meanings given to those processes. The interpretations made by different cultures, including our own, is at the root of change and of our understandings about international life.

Alexander Wendt has skillfully outlined the epistemological problems in the levels-of-analysis debate which I will not reiterate; suffice it to say that what he has done is important and of interest to me here because it begins to establish a third theoretical perspective that he calls constructivism.[5] The purpose for defining this constructivist position is to highlight what is minimized in neo-realist and neoliberal literature on international cooperation, "namely the issue of identity and interest formation."[6] Wendt's point is that the neorealist's (Waltz) focus on power and human nature precludes cooperation, and neoliberals (Keohane) argue that knowledge and institutions make it possible. According to him, what both schools of thought overlook by bracketing a state's identities and interests as assumptions are possibilities and/or strategies for cooperation, as well as misrepresenting the dynamics of cooperation.[7] "Both treat interaction as

---

5. Alexander Wendt, "The Agent-structure Problem in International Relations Theory," *International Organization* 41, no. 3 (Summer 1987): 335-370; "Anarchy is What States Make of It: the Social Construction of Power Politics," *International Organization* 46, no. 2 (Spring 1992): 391-425; "Collective Identity Formation and the International State," *American Political Science Review* 88, no. 2 (June 1994): 384-396; "Hierarchy under Anarchy: Informal Empire and the East German State," *International Organization* 49, no. 4 (Autumn 1995): 689-721.

6. Wendt, "Anarchy is What States Make of It," 393.

7. Wendt, "Collective Identity Formation and the International State," 384.

affecting only the price of behavior and thus assume that a rationalist research agenda exhausts the scope of systematic theory."[8]

Wendt points out that "states are engaging in discursive practices designed to express and/or to change ideas about who 'the self' of self-interested collective action is."[9] I would challenge Wendt to take this a step further by saying that this is not only about the presentation of 'self' in international discourse, but it is about interpreted meaning. What does cooperation as a concept mean? Wendt himself uses existing academic definitions of cooperation and overlooks the possibility that cooperation may have different cultural interpretations. Not only is anarchy what states make of it, but so is every other concept that American analysts use to make sense of a state's behavior. The important question that I want to ask in this book is, "how do our concepts of cooperation and thus transformation change when we consider that state's identities shape the interpretations made of interaction?" This requires a reformulation of the levels-of-analysis gap from a problem of how to best conceptualize what is observed to one of asking how our primary analytical concepts must change when we consider alternative explanations or interpretations. It is precisely because neorealism and neoliberalism have bracketed states' identities as external factors to their analysis that the levels-of-analysis gap exists, and as Wendt rightly points out this is a mistake.

The issue that Wendt overlooks and which I want to take up is the question of interpreted meaning. For this reason I hope to add to Wendt's creation of a constructivist theoretical position by using this book's analysis of Chinese domestic politics as a means of commenting on the prevailing notions of international cooperation. In doing so, I am using domestic politics to inform international analysis and so in a different way I am crossing the levels-of-

---

8. Ibid., 391.

9. Ibid.

analysis gap. I support the general position articulated by Alexander Wendt, but believe that the hermeneutical approach needs to be maintained as a valid means of inquiry. Hermeneutics underlies the theory of structuration as outlined by Anthony Giddens and must be, as Giddens states, one of the primary tasks of sociological analysis.[10] Wendt must consider this as well if he is to continue to advocate "sociologies of international relations" based upon an agent-structure epistemology.[11]

To bridge the micro/macro 'gap,' the problem is in need of redefinition. As the problem now stands the question to answer is, how are states influenced by and reciprocally influence the international system? The assumption behind this question is that the analysis of domestic politics and culture obfuscates generalization and analysis at the international level, whereas the theoretical generalities at the international level cannot adequately deal with the influences of domestic politics. With this question the analyst/observer and their definitional constructs are separate from that which is being observed, and the struggle is to find the 'right way' to define and analyze concepts such as international processes, actors, change, cooperation, learning, power, etc. All of these concepts are vigorously debated in an attempt to understand the relationship between states and their international relations.[12] What is not considered in this debate is the role of the analysts themselves and their assumptions in shaping culturally and politically specific definitions of international politics.

The reformulation of the problem that I propose is to see the 'gap' as one between our own cultural proclivities and other meanings given to what we

---

10. Anthony Giddens, *New Rules of Sociological Method: A Postive Critque of Interpretive Sociologies* (Stanford: Stanford University Press, 1993): 170.

11. Wendt, "Hierarchy Under Anarchy: Informal Empire and the East German State," 691-95.

12. For summaries of these arguments and innovative conclusions see: Alexander Wendt, "Levels of Analysis vs. Agents and Structures," in *Review of International Studies* 18 (1992), 181-5; Wendt, "The Agent-Structure Problem in International Relations Theory," *International Organization* 41, no. 3 (Summer 1987), 335-70; James N. Rosenau, *Turbulence in World Politics: A Theory of Change and Continuity*, (Princeton: Princeton University Press, 1990).

observe. By using an analyses of domestic politics geared towards understanding political cues and identities so as to understand other culturally significant symbolic realities and to then use these concepts to make interpretations of international structures, such as the UN, the analyst's assumptions and concepts can broaden to become more inclusive and insightful. The 'gap' is between American cultural assumptions of meaning and other domestic or local interpretations of meaning. The problem is that in order to better define and conceptualize what is observed, the analyst's own assumptions which inform the concepts used need to be challenged by including other perspectives. If anarchy and cooperation are what states make of it, the obvious question is "what do they make of it? What does it mean?" Anarchy and cooperation can only have the power to shape state's interests and identities by virtue of the interpretations that state's leaders make. Anarchy and cooperation however are concepts from the social sciences, not issues debated by foreign governments. Wendt gives these concepts the power to transform people's political interests and identities thereby maintaining the cultural primacy they have for American international relations theorists, but it is only by virture of how people in other societies understand and give meaning to these concepts, if any at all, that change occurs. By using the culturally distinctive concept of model-following as a domestically based articulation of Chinese social structure and power to inform my interpretations of Chinese international behavior I address the levels-of-analysis problem as I have redefined it, as an issue of cultural interpretation.

To illustrate this point more fully I will attempt to use Waltz's neorealist definitions, which are still the foundation stones for international relations theory, to illustrate the interpreted nature of the international system, as a symbolic reality that is defined through language and open to interpretation and multiple meanings. I will then provide an illustrative, alternative explanation of Chinese behavior in the United Nations based upon the concept of model-following behavior learned from the analysis of ecological agriculture. By looking at this case study of Chinese participation in the UN I am focusing on an application of the theories of international cooperation, and in doing so I hope to provide both a critique and a rethinking of the idea of cooperation to stimulate new questions for political analysis.

International Systems, Cooperation, Interpretation

"The structure of a system is generated by the interactions of its principal parts." Systems theory then is a study of "states that make the most difference"; it is "necessarily based upon the great powers." Waltz recognizes that structures do not act, but they do affect the behavior of actors within the system "through socialization of the actors and through competition." Socialization "limits and molds behavior" and "encourages similarities of attributes and of behavior....Competition generates an order, the units of which adjust their relations through their autonomous decisions and acts." A theory of structure is about positioning components. In the international system states, as entities which are not bound by a single government, are largely autonomous from one another. This situation creates an anarchical arrangement where the distribution of capabilities creates strong and weak powers.[13] According to Waltz it is emphatically not about how states interact or relate to one another, yet he proceeds to describe a system that is a distinctly a product of state's interaction.

Relying on economic theory for his own interpretation, Waltz's frequent analogy for the international system is the market place. "Once formed, a market becomes a force in itself, and a force that the constitutive units acting singly or in small numbers cannot control." In the same way the international system becomes the 'invisible hand' of state behavior. It is by each individual state following self-serving interests in competition that the "greater good of society is

---

13. Kenneth Waltz, "Reductionist and Systemic Theories," in Robert O. Keohane, ed., *Neorealism and Its Critics*, (New York: Columbia University Press, 1986), 61-65. Waltz does acknowledge that the study of states behavior is important work, but it is descriptive and reductionist and can not be used to generate theories of the international system, which is the anarchical organization of states.

produced." The international system is thus "formed and maintained on a principle of self-help [individualism and competition] that applies to the units [states]." Domestically the self-help principle "applies within governmentally contrived limits," but international politics is less guided. The goal of states existing in this market-like international system is survival. "Survival is a prerequisite to achieving any goals that states may have, other than the goal of promoting their own disappearance as political entities." The ability to achieve the goal of survival is a function of state's political, economic, and military capabilities. Within the international structure the distribution of capabilities is uneven and results in the observation that there are strong states and weak states. In the end "international political structure (is defined), by counting states. In the counting, distinctions are made only according to capabilities."[14] The definition of structure that this counting leads to is this; the international structure is made up of the formal and informal rules created and enforced by states through their competitive interaction which is the result of following self-serving interests. These rules are expressed through treaties, regimes, international law, international organizations, diplomatic understandings, etc. The corollary to this definition is that stronger states will have greater ability to define and enforce the rules that make up structure and so influence the nature of inter-state relations. Having defined the international structure Waltz goes on to explain how this structure affects state behavior.

What are the operational characteristics of Waltz's definition of structure used as an analytical concept? An international structure, defined as rules, works to force states to consider the results of their actions in terms of the consequences those acts may or may not have for their survival. Rules, whether they are understood tacitly or explicitly expressed in documents, give government leaders a sense of consequence to their self-serving actions that forces a form of systemic constraint on interstate cooperation.

---

14. Waltz, "Political Structures," in *Neorealism*, 83-94.

> A state worries about a division of possible gains that may favor
> others more than itself. That is the first way in which the structure
> of international politics limits the cooperation of states. A state
> also worries lest it become dependent on others through
> cooperative endeavors and exchanges of goods and services. That
> is the second way in which the structure of international politics
> limits the cooperation of states.[15]

Because states as individuals are self-serving in their struggle for survival they avoid cooperative actions that will detract from their own advancement. In this way "structures encourage certain behaviors and penalize those who do not respond to the encouragement."[16] This is the socializing effect of structure upon state behavior, a way of bringing states into a shared understanding, a shared interpretation of what 'acceptable behavior' is or is not.

According to Waltz structures, as rules may change in one of two ways, either by "changing the distribution of capabilities across units" or by "imposing requirements where previously people had to decide for themselves." Both means of change are changes in the rules; the first is a change in who is capable of making rules and the second is the creation and enforcement of new rules. For example,

> if some merchants sell on Sunday, others may have to do so in
> order to remain competitive even though most prefer a six-day
> week. Most are able to do as they please only if all are required to
> keep comparable hours. The only remedies for strong structural
> effects are structural changes.[17]

It is from Waltz's definition of rules of international play as structure that balance-of-power theories emerge to explain state behavior. In balance-of-power theories, states seeking their own survival as a minimum goal use either "internal

---

15. Waltz, "Anarchic Orders and Balances of Power" in *Neorealism*, 103.

16. Ibid.

17. Ibid.

efforts (to increase economic capability, military strength, develop clever strategies) and external efforts (moves to strengthen and enlarge one's own alliance or to weaken and shrink an opposing one)." This requires at least two states or 'players,' two powerful states with advantages in capabilities that will attempt to find balance with other comparable powers or competitors for resources and allies. "Great tasks can be accomplished only by agents of great capability. That is why states, and especially the major ones, are called on to do what is necessary for the world's survival."[18] In this 'king-of-the-hill' approach to international politics, the golden rule is that the one with all the gold makes the rules.

In such an international system the creation and enforcement of rules do not affect all equally. Waltz presents this inequality clearly in his discussion of competition as a desirable condition. Competition is desirable because it leads to sameness. In contrast to the idea of competition is the concept of socialization, which Waltz defines as a half-hearted conformity to international rules. This demarcation of insiders and outsiders, full players and unwilling participants, implies the existence of different interpretations given to the rules by different state governments. The leaders of strong states may create rules that they perceive are necessary for survival, but these same rules will not be interpreted similarly by all global participants. Rules are as exclusive as they are inclusive; they symbolize one social group's vision of what is real and necessary for survival.

According to Waltz, following the rules has the effect of socializing states to what the rules define as appropriate behavior, rules that are created by the strong states, using the language of those same strong states, embodies the values of those states. In this system of survival through competition, weaker states seeing the success of the stronger states and interpreting this success as good, will "emulate them or fall by the wayside." The two operating principles of the

---

18. Ibid., 107-117.

international system, competition and socialization enforce this 'emulating' behavior. "Competition produces a tendency toward the sameness of competitors," best exemplified by the Cold War arms race. Socialization is a "conforming to common international practices even though for internal reasons they [states] would prefer not to."[19] While Waltz makes a definitional distinction between competition and socialization as degrees of conformity, both are forms of socialization more broadly defined as learning how to get along according to a predominante set of rules for 'acceptable behavior.'

For Waltz, competition occurs among conforming states and leads to emulation and sameness, whereas socialization is the appearance of conformity by nonconforming states. Socialization is an important concept because it refers to how states interpret the rules of play to best meet their interests, rules that most states have not had a hand in creating. Nonconforming states are those which largely remain outside the rules of competition and will not emulate and therefore remain different. These states are a potential challenge to the status quo of competition because they do not accept the rules but conform only out of necessity. Conforming is good because competition is good. It is a means to achieving ends, and nonconformity is bad because it is a challenge or potential challenge to the rules. Socialization does not always require the presence of a hegemon or of a great power, as Waltz claims, to create and enforce the rules. The socialization of nonconformist states proceeds at a pace that is set by the extent of their involvement in the system. And that is another testable statement."[20] It is this 'testable' statement, how to involve nonconformist states in the system so they will want to compete according to the rules, that is pursued by Robert Keohane in his theorizing about international cooperation.

---

19. Ibid.

20. Ibid., 128-29.

Understanding cooperation in international affairs is important because it is through cooperation that international players come to compete according to agreed upon rules, and through 'friendly' competition can work towards the Waltzian goal of survival. Keohane questions Waltz's belief that there needs to be a strong state to do what is necessary for the world's survival,' where world survival becomes a metaphor for the survival of the strong state. Robert Keohane elaborates his theory of cooperation, with the question of "how can order be created out of anarchy without superordinate power; how can peaceful change occur?"[21]

To understand this a clear definition of cooperation is necessary. Cooperation is not harmony;

> harmony refers to a situation in which actor's policies (pursued in their own self-interest without regard for others) automatically facilitate the attainment of other's goals....Cooperation requires that the actions of separate individuals or organizations - which are not in pre-existent harmony - be brought into conformity with one another through a process of negotiation, which is often referred to as 'policy-coordination.'[22]

Cooperation then is the result of policy negotiation which is necessary to avoid or minimize conflict among states as they pursue individual goals. "Cooperation is highly political: somehow, patterns of behavior must be altered. This change may be accomplished through negative as well as positive inducements." The most successful strategies for "attaining cooperative outcomes" are those that combine "threats and punishments as well as promises and rewards."[23] The question of power here lies with strong states and their ability and willingness to "maintain the essential rules governing interstate

---

21. Robert O. Keohane, "Theory of World Politics," in *Neorealism*, 199.

22. Robert O. Keohane, *After Hegemony: Cooperation and Discord in the World Political Economy*, (Princeton: Princeton University Press, 1984), 51.

23. Ibid., 53.

relations."[24]   In a world without such a hegemon, it is the rules embodied in international regimes such as the North American Free Trade Agreement (NAFTA) that "can be important symbols that legitimize cooperation or as guidelines for it.   But cooperation, which involves mutual adjustment of the policies of independent actors, is not enforced by hierarchical authority."[25]

> Keohane's definition of cooperation as being embodied in the rules of international regimes is a modification of Waltz's understanding of system change.   International regimes and their organizations can affect the distribution of state's capabilities through the imposition of rules that guide action and decision making where previously state leaders were left to their own judgment.   By facilitating the transmission of information, international organizations reduce uncertainties and asymmetries of access to information.   By providing a setting for regular meetings of members, international institutions affect transaction costs by 'making it easier for governments to get together to negotiate the agreements.' Once agreements are in place, acts that violate them have more serious implications than would have been the case in the absence of these agreements. Retaliation may be justified, and violation of an agreement with respect to one issue may call into question the sanctity of agreement concerning other issues....regimes promote compliance with their rules by affecting "the calculations of self-interest in which rational, egoistic governments engage."....A state's failure to comply with one rule may affect the future willingness of other states to follow this and other rules.[26]

What emerges here is a definition of power that is based upon a state or nonstate international actor's power to alter existing rules or create and implement new rules.   Power is also a function of being successful, thereby defining the direction and nature of the competition, so that other states will want to 'emulate' the strong state(s).   What is 'successful' and what rules are determined necessary

---

24. Ibid., 34.

25. Ibid., 237.

26. Robert Keohane, *After Hegemony* in Jacobsen and Oksenberg, 7.

and the transformation of rules are all subjective acts based upon people's interpretations of what is and will be necessary to maintain and achieve the subjectively defined conditions that will permit survival. Judgment and decision making are removed from individual leaders acting in isolation of one another, and through negotiation an agreed to set of rules are created.

In this capacity regimes have the power to alter the international structure through the process of socializing state political and economic leaders to the currently established rules of international play. Conflict that may occur through open competition is limited through policy negotiation, as, for example, the General Agreement on Trades and Tariffs (GATT) is intended to do. Enforcing/ following (written this way to emphasizes the power relations inherent in rules) a set of rules for expected international behavior socializes by rewarding conformity. Following rules is a form of ritual behavior that connects individual action to the larger social group and that socializes the individual. The 'homogenizing' (to use Van Ness's term) effect of competition and the modification of behavior or socialization are best achieved by using 'threats, punishments, promises, and rewards.'

Using Waltz's example of the store owner who is willing to stay open on Sundays, one sees that it is this individual act that results in system change and alters the rules of play for the other store owners. The innovative store owner can either be seen by the other store owners as a 'nonconformist' or as a 'strong' leader. Waltz implies that the innovative store owner is a leader because this person's potential level of income; therefore success has been increased and the other store owners must react by emulating the now more successful leader to stay competitive. The new systemic rule of being able to work Sundays changes the norms of behavior and resocializes the other owners. If the innovative store owner is viewed as a nonconformist by the other stores, they may act collectively with punishments, threats, promises, and rewards to force the rogue store owner to stay closed all or partially on Sundays: cooperation through policy coordination.

Power lies in the hands of those able to define and enforce the rules of competition and of socialization. What if it is against a store owner's beliefs to be open on Sundays, but they are competitively forced to conform. One may perceive compliance with the rules, apparent cooperation as a result of competitive necessity, but one does not see or understand how an alternative interpretation of the rules may affect the process/system relationship through changes in the rules or new rules created at a later time. Rules are based upon culturally specific definitions of idealized behavior that are the products of power holders and that require enforcement. Homogenization, socialization, conformity, competition are terms that have negative and positive connotations; they create spaces of inclusion and boundaries of exclusion.

This seems rather basic, yet as Americans, members of a strong state that creates, influences, and enforces the rules for international play, our theoretical assumptions and the analytical concepts we use to interpret the world's political activities do not question the rules and the multiple meanings that those rules may engender. There is a tendency to only view the positive, idealized measure of behavior as an appropriate political objective with positive or negative results based upon only two choices: to conform or not to conform. Conflicting interpretations over the rules and the actions they authorize become conflicts over whose interpretation is 'right' rather than a source of questions about how these multiple interpretations are facilitated and how they can be included in a dialogue about the rules.

We see this sense of either 'for us' or 'against us' when Waltz notes the Bolsheviks unwillingness to be socialized and comments that "refusal to play the political game may risk one's own destruction."[27] This same idea is more subtly presented by Jacobson and Oksenberg when they note that the IBRD

---

27. Waltz, "Anarchic Orders and Balances of Power," in *Neorealism*, 128.

(International Bank for Reconstruction and Development, or the World Bank), IMF, and GATT have not "converted China to capitalism nor subverted the international institutions from their basically neoliberal orientation nor led the United States astray."   China is in a position of having been converted to capitalism or not, and the international institutions are also either altered or not. There is no alternative possibility, the focus is on the stability of the status quo as defined by the United States as a strong state.   This conclusion is intended to show the politically noncoercive nature of international organizations and how contact with the IBRD, the IMF, and GATT have led to institutional changes within China to facilitate "information exchange and social learning."[28]

Information exchange and social learning is directed towards the Chinese as a means of socializing them to international practice.   In the case of the World Bank, learning takes place through the process of loan negotiation and compliance.   The stipulation of marketing, resource access and distribution, infrastructure, pricing, and accounting requirements for a successful project and then the enforcement of these requirements as negotiated conditions of a loan/project are the basis of social learning.   World Bank loan officers recognize that it is the day-to-day, rule-by-rule changes that make up the process of socialization to different norms.   Instituting a new accounting practice, or ensuring the 'correct' interpretation of a contractual point are ways of not only teaching specific practices but also enforcing the values embodied in the language of those practices.

The effect of the international system on states is always and in all cases the product of interpretation, interpretation by the states and nonstate actors of what their relationship is to the system and of what the perceived effects of it are. What meaning the people that represent nations give to competition, conformity,

---

28. Harold Jacobson and Michel Oksenberg, *China's Participation in the IMF,the World Bank, and GATT: Toward a Global Economic Order* (Ann Arbor: University of Michigan Press, 1990), 159-60.

cooperation, and interdependence (as a form of placing conforming limits on competition) has ramifications for future actions that will change the structure of the system. International actors create the system by virtue of their actions, change it by virtue of their actions; at all times these actions are based upon interpreting the meanings of the system/rules to gauge their next policy decision, action authorized by interpreted meaning.

To understand the effects of structure on process must be an interpreted understanding. If Waltz and Keohane use concepts inspired by economic theory to provide one interpretation of the international political and economic structure, is it not also appropriate to ask how those involved in the process interpret the structure? What meaning do they give to the rules and how do the actions determined by these interpretations alter the process and then again the structure? Learning, following, and understanding rules may lead to a form of interaction perceived as socialization, but it also facilitates the creative power to change the rules and to redefine conformity through a reinterpretation of the rules.

Regimes and their organizations are the products of people coming together to create rules that they believe are beneficial generally and/or especially to their advantage. The rules become 'symbols' of cooperation, of norms for socialization, generators of spheres of inclusion/exclusion, and justifications for forms of power. As symbols these rules are open to multiple interpretations, interpretations based on a variety of historical and cultural backgrounds and experiences. The process of socialization or forced exclusion on someone else's terms is not always pleasant or successful and will eventually result in a policy reaction, or challenge to the existing norms, a challenge that if successful can alter the system. This realization by the rule creators can be seen in the following comment about China's participation in the IBRD, IMF, and GATT. The uncertainty of China's ongoing socialization points to the need for these international organizations to remain "vigilant, agile, and bold in defense and

210

pursuit of their (neoliberal) mandates and values if they are to remain a vital force."[29]

This call for the strong member nations, primarily the G-7 (the United States, Canada, France, England, Italy, Germany, Japan) which are creating rules and enforcing them, to remain 'vigilant' belies the image of political neutrality often given to international organizations. Jacobson and Oksenberg, expressing this perception of neutrality, see the IMF, IBRD, and GATT as international organizations that "are instruments for achieving noncoerced cooperation among sovereign states through the collection and dissemination of information and by providing a forum for reaching consensus."[30] Yet they notice that China's admission to these institutions could not have occurred without the support of the United States and that "the United States had little impact on the day-to-day interaction between China and the IMF and the World Bank other than its broad influence in determining the level of funds available to the two institutions."[31] Again their point is to emphasize the neutral aspect of international institutions, yet the 'broad influence' of the United States in determining levels of funding is hardly a neutral act.

By virtue of the language embodied in the rules of a given regime, the values and goals of its creators will be ever present. Regime viability does not require the direct control of daily operations by a strong power. An example of the IBRD loan officer's daily dealings with the Chinese reiterate this point. In a telephone interview I was told, in response to my question about American government influence in his decisions about loans with the People's Republic of China, that he operated as an officer of an international organization and not as a representative of the United States or at the specific direction of the United States.

---

29. Ibid, 159-60.

30. Ibid, 159.

31. Ibid. Emphasis added.

Instead he is an employee of an international institution bound by the rules created by the member states.[32] This organizational autonomy gives the illusion of neutrality and independence, belying the political and economic influence of strong states to maintain a 'proper' interpretation of rules. It is similar to a foundation with a rich sponsor who defines how the monies are to be spent. The sponsor can be uninvolved in the daily workings of the foundation; it is the prospect that if the rules are not followed the sponsor will withdraw support that maintains compliance by the foundation administrators.

The concepts of interdependence and democratization as American analysts define them should not be limited in consideration as ideal end goals to be achieved, nor can the concepts of cooperation and liberalism be limited to measuring the level of interdependence and democratization to date. Instead interdependence and democratization are predominate concepts relevant to the twentieth century, they are a moment in the process of global social evolution. These concepts are culturally ours and because America has been a great power since WWI, Americans have been very influential in defining the current rules of interaction used in the international system. As Americans we are not, however, appreciating the various interpretations that other cultures may be giving to these rules and their concomitant socialization. How different societies interpret the processes of change as they impact them at local levels will have ramifications for their interaction with the process at international levels. These are the interpretations that will affect future changes in the international system, a system defined broadly as the political and economic rules of the game, rules delineated by strong state's behavior and the rules embodied in regimes and enforced by organizations.

---

32. World Bank loan officer, China Division, interview by author, August 1993.

Model-Following as an Alternate Interpretation

Ecological agriculture is in part a product of China's interaction with the process of global change, a reaction to excessive reliance on external factors and influences, a reaction to the socialization process, and it is being translated into local action. Its expression and implementation are based on symbolic language that is culturally relevant and politically meaningful; it is symbolic of a part of Chinese identity that is powerful enough to develop and implement a national policy. Having outlined the themes of model-following in Chapters 4 and 5, the concept can now be appropriated and used independently of the domestic context within which it was analyzed. The concept of model-following embodies a specifically Chinese understanding of social structure and of power relationships, and the question is now one of "what are the implications of this concept for understanding Chinese international behavior?"

American analysts tend to see China's increasing competition in the world and cooperation in international organizations as evidence of peaceful integration in the world system, the socialization of a communist, nonconformist regime into a potentially democratic state. How must one's concepts and conclusions change if one asks, "how does the concept of model-following as defined in this book, and found to be a component of Chinese self-identity, shape the meanings that are given to themselves as a culture and to their international actions?" I will first apply this concept to Samuel Kim's analysis of China's participation in the United Nations, providing an alternative explanation and conclusion; then as a means of summarization I will also apply this concept to Waltz's analogy of a store owner staying open on Sundays.

Samuel Kim's analysis of Chinese behavior in the Security Council makes the following observation:

> The continuous operation of UN (peace keeping missions) shows
> that Chinese ideological opposition to peace-keeping expressed in
> the form of non-participation and dissociation translates into

> Chinese 'cooperation' in the form of noninterference in the [UN]
> authorization and implementation process.[33]

In addition to Chinese cooperation through noninterference, Kim also observes that the Chinese seem to be less concerned with the legal aspects of any given issue. He cites the Chinese attack on the Soviet abuse of the veto but only mild regrets when the French vetoed admission of the Comoros in 1976.

> At times the Chinese diplomacy in the Security Council has conveyed the impression of following an actor-oriented rather than an issue oriented approach....The Chinese sometimes showed more interest and concern about the who question than about the what question....Apparently, the Security Council is valued more as a sword with which to fight against the revisionist social imperialists than as a shield to protect the interests of the Third World.[34]

Kim is concerned about the apparent incongruity between what he labels as Chinese verbal rhetoric and observable actions taken. He gives the following example. During Angola's struggle for independence from Portugal three liberation groups came into play; (1) the Popular Movement for the Liberation of Angola (MPLA), (2) the National Front for the Liberation of Angola (FNLA), and (3) the National Union for the Total Independence of Angola (UNITA). China preferred UNITA but supported the Western-backed FNLA against the Soviet-backed MPLA. Responding to the Organization for African Unity's (OAU) call for neutrality among the three factions, the Chinese removed military advisors from the FNLA camps. After the MPLA won the civil war, China refused to extend diplomatic recognition to the new Angolan government. Additionally, China opposed through non-participation Security Council Resolution 387 that condemned South African aggression against Angola and called for South Africa to compensate Angola for the damage and destruction of the war.

---

33. Samuel Kim, *China, the United Nations, and World Order* (Princeton: Princeton University Press,1979), 238-41.

34. Ibid., 239.

214

In Kim's analysis this reveals incongruities between China's principles of being a leader against anticolonial forces and of supporting the victims of aggression. First, China chose an obviously anti-Soviet stance, siding with the United States and against China's own preference. Second, China did not extend recognition to a new government because it was Soviet backed, and lastly, believing that condemnation and compensation should be the responsibility of all countries with foreign military forces in Angola (the USSR and Cuba), China sides with the United States, France, Japan, Italy and England in abstaining from the vote on resolution 387. For Kim this reflects an inconsistent stand between the verbalization of principles that support the victims of colonial aggression and the pursuit of national interests by nonparticipation in Security Council voting.

Referring to China's predisposition to abstain rather than veto a Security Council resolution, he asks,

> Would it not make more sense to veto a scrap of paper than to attack it after it had already been adopted? This is indeed the have your cake and eat it too posture that China has adopted in the Security Council. It is the Chinese way of reconciling principles and interests.[35]

Again, Kim is speaking to the seemingly contradictory way, a picking and choosing by the Chinese of arguments and issues that are not consistent with his interpretation of Chinese principles. He observes that China opposes peace keeping operations based upon closed door consultations between super powers.[36] The Chinese instead believe that the "dispute should be settled in a reasonable way by the countries concerned through consultation on an equal footing."[37] For Kim China's failure to actively oppose that which it believes is wrong is a signal

---

35. Ibid.

36. Ibid., 214.

37. UN Doc. S/PV.1612 (13 December 1971), 1-2, in Kim, 217.

of the gap between rhetoric and passive support and is not the action that a strong leader would take.

> China's opposition is more verbal than real. She cast no negative vote on such instrumentalities of the Security Council as the United Nations Forces in Cyprus,....in the Middle East,....the Rhodesian Sanctions Committee. As a result the above mentioned operations continued unhampered.[38]

About China's Security Council involvement, he concludes that "China has maintained a posture of mutual respect and peaceful coexistence, and....that China has shown no sign of disinterest or disillusionment with the Security Council is a hopeful sign."[39] For Kim, China is following the rules and is therefore cooperating with the other security council members, and this is a "hopeful sign" that China is not a radical, revolutionary country intent on using the U.N. for furthering its own communist agenda. China's use of nonparticipation and abstention from Security Council voting are viewed in Kim's analysis as passive acts of cooperation as opposed to active acts of obstruction, acts that try to balance what he labels as rhetoric against national interests.

> Caught between principles and interests, Chinese action often seems to follow the latter, whereas Chinese rhetoric pays lip service to the former. China has often escaped the burden of taking a clear and consistent stand on peace-keeping operations by opting for nonparticipation.[40]

As an interesting and complementary note, the author of a study of China's Security Council participation during the 1990 Gulf War led to the following comments.

> All activities against Iraq were coordinated and supervised by the United Nations through several Security Council resolutions. It

---

38. Kim, 217.

39. Ibid., 241.

40. Ibid., 229.

was recognized that China played an instrumental role in reaching these resolutions, sometimes, as in the case of Resolution 678 [authorizing use of force in Kuwait, and which China abstained from voting on], a crucial role. Indeed, China's vote in the UN Security Council has been a principal source of its new international influence. Much of China's influence is, however, still an illusion because the PRC did not have much choice but to support, or at least not oppose, the Security Council's resolutions.[41]

The conclusion arrived at is that by not participating China did not obstruct and that therefore China is not a strong power in world affairs. China took minimal risks to again balance their principles and to conform to the needs of their own national interests. "Sitting on the fence and watching the tiger fight, China still plays a marginal role in world affairs."[42] It is this very 'fence sitting' strategy that Barbara Tuckman observes as used by the Chinese during World War II as Nationalist and Communist Chinese leaders let the Americans and British fight the Japanese, saving their own strength for the inevitable civil war that would follow.[43] Two generalizations can be made here about a Chinese sense of political power: (1) power is not always about overtly getting people to do what one wants them to do, and (2) the uses of power do not need to be showy displays. The questions that I want to address by using model-following for a reexamination of Kim's analysis are to what degree is there a 'gap' between Chinese verbal and behavioral rhetoric and when is nonparticipation not cooperation?

The key components of Samuel Kim's analysis are that China exercises nonparticipation rather than veto as a negative response to Security Council initiatives and second that China is more concerned with who is involved in an

---

41. Yitzak Shichor, "China and the Gulf Crisis: Escape from Predicaments," *Problems of Communism* 40, no. 6 (Nov/Dec. 1991), 85. Parenthetical note added for clarification.

42. Ibid., 90.

43. Barbara W. Tuchman, *Stilwell and the American Experience in China: 1911-1945* (New York: Bantam Books, 1971).

international incident rather than the legal question of what is involved. Kim sees this as contrary to China's claimed leadership position as protector of the Third World. Third, and a point that is related to one and two, China's voting actions and verbal actions seem to be contradictory in an attempt to obstruct neither superpower nor Third World interests. Kim seems to imply that this is a passive, and less valuable path to follow in light of China's claims to being a Third World leader.

As presented by Kim, these evaluations indicate that Chinese behavior in the Security Council has been largely symbolic and tactical. Kim also indicates that protracted involvement with the UN has the potential to change further Chinese behavior and attitudes. These conclusions deflate China as a power of any significant threat and reiterate the point that Chinese exposure to Western liberalism through continued participation in the United Nations is a major force for change in China, both of which place Western values and our concerns at the center of the analytical spot light and ignore what possible interpretations the Chinese might be giving to their own experiences.[44] To ignore the perceived 'gap' between verbal actions and observed behavior is to maintain an observer based definitions of leadership, and power.

How does Chinese model-following as a cultural theme explain the same situations which Kim has explained, and what are the implications of this alternate, cultural explanation? First, China's perception of itself as a world power in the Security Council must be understood. China's concept of a superpower is defined and based upon how different understandings of how a strong nation should act. This is labeled by Kim as a behavioral definition and lies in contrast to defining nations, as Waltz does, as solely based upon military and economic capabilities, with the interpretation of the international structure by

---

44. These conclusions do not differ significantly from Jacobson and Oksenberg's. Whereas Kim examines specific behaviors in the UN, the other authors consider the process of accession as indicators of rule acceptance and following: conformity.

prevailing strong powers, primarily the G-7, as the 'correct' interpretation.[45]   I will use this label of 'behavioral' to maintain the distinction between Chinese and Western interpretations, but it must be understood that a Waltzian perspective is also behavioral albeit a different measure of acceptable behavior based upon a different understanding of the rules and the sense of structure and power they imply. A behavioral definition is one based on actions, on defining and following different interpretations of 'correct' behavior. For the Chinese this does not minimize the importance of having strong capabilities, but the emphasis shifts from a Waltzian perspective of potential use through capabilities to a definition and measure of actual behavior: how the potential is used. For the Chinese the behavior of the United States is percieved to be hegemonic, coercive, chauvinistic, pursuing policies of monopoly and is therefore not a superpower.

China perceives itself to be a superpower by adhering to the Principles of Peaceful Coexistence created at the 1955 Bandung conference. These principles are respect for other's territorial integrity, nonaggression, noninterference in other's internal affairs, equality and mutual benefit, and peaceful coexistence. One may quickly point to China's violations of these principles, but the argument is one of defining the concepts of peaceful coexistence, of 'right' behavior and whether one uses one's own or a Chinese definition. For example Americans tend to see China's desire to reunite with Taiwan or occupation of Tibet as a lack of respect for other's territorial integrity and interference in their internal affairs, but from a Chinese definition these are both historically territories of China and the principles are not being violated.[46]

A Chinese-based behavioral definition of a superpower is also supported by the model-following concept through the notion of paternalistic leadership.

---

45. Samuel Kim, "Behavioral Dimensions of Chinese Multilateral Diplomacy," *The China Quarterly* 72, (December 1977), 713.

46. In what ways are our own continuing territorial disputes with Native Americans reflected in this argument?

This is leadership that provides an exemplary model of behavior from within a Chinese conception of what is exemplary for others to emulate. It is based on the paternalistic power, the principle of having correct or beneficial knowledge and the ability to put successfully into practice that knowledge so as to facilitate its transfers by example and teaching. Following a model is a function of realizing and accepting the differential nature of social relations and of the responsibilities people have in those relationships. Being a leader and therefore a model for others requires continual self-improvement to maintain the status of a model and to emanate the virtues that will stimulate others to learn from your example. Following a model also requires continual self-improvement to achieve and eventually surpass the standards of the model, becoming a model in one's own right.

China's self-perception as a leader and its interaction within the group situation of the Security Council can be seen as being related to the concept defined in this book as model-following. China's behavior is symbolic as Kim suggests, but not in the hollow way that he implies. Rather, China's behavior in international organization is symbolic of the perceptions the Chinese government representatives have of themselves relative to other nations, an understanding of the nature of the relationships of which China is a part, and of what constitutes appropriate behavior in those relationships, as well as interpretations and uses of power.

China's role as a model leader is to act in a manner fitting a world leader within an international organization, such as the U.N., which ideally is intended to transcend nationalist politics. Coercive exercises of power are not becoming of such a leader. Acting in small group settings prior to a Security Council decision, the Chinese have no qualms about making their positions clear. However, within the context of Security Council voting, abstention from voting rather than a veto can be seen as a strong indication of China's opposition and at the same time a desire not to be disruptive of the processes of the Security counsel and of not wanting to been perceived as being coercive in building coalitions as the United States is perceived to be. As Kim observed the Chinese do not like America's predilection for closed door negations between a few key powers. Building coalitions is a form of individualism that places the interests of one above the interests of the majority. As Lucian Pye points out, within Chinese culture "there is a tendency to see individualism in a negative light, and as the source of corruption and selfishness."[47]

Within a model-following-based definition of leadership, a model leader maintains unity and supports the general welfare. If one is an exemplary model, others will see the correctness, or the virtue, of that leader's position and want to voluntarily emulate the spirit of the model. Kim may see Chinese nonparticipation as cooperation, but for the Chinese it may signal a strong leadership stance, not necessarily a cooperative action. Nonobstructive behavior and passive protest is markedly different and implies more sophisticated forms of interaction than a definition of cooperative behavior that is based only on negotiated rule formation and rule following.

The observation that the Chinese are concerned with who is involved in an international incident, rather than the legal question of what is involved, is not contradictory to behaving as a model and valuing paternalism as a power base. Kim sees the focus on who, in this case the USSR and the USA as imperialist nations, as contrary to China's position as protector of the Third World. China's focus on the who, the United States and the former Soviet Union, as imperialist is an expression of their interpretation the Third World plight as the pawns of imperialist advances. By focusing on the perpetrators behind an event, China is looking beyond the legal imperatives of a specific incident to the power relationships that facilitated the incident in the first place. As a paternalistic leader, China's focus on the larger powers reenforces its own position as a model

---

47. Lucian Pye, *The Mandarin and the Cadre* (Ann Arbor: University of Michigan Press, 1988), 60.

leader working for the general welfare of all nations. The focus on who is involved rather than the legalistic aspects also reflects a historical cultural preference to human relations over the legal contracts which are themselves a product of existing sets of social/power relations. So to argue a legal point is to work within the definitions created by existing power relationships dominant in any given issue area. This means that depending upon who is involved and the rules and definitions in play, the Chinese may or may not acknowledge those rules as legitimate or binding upon them. To focus on the power basis of relationships is to establish and/or maintain sets of relationships and define one's position relative to them (also a part of defining self in terms of others and selecting models to follow). This allows China to claim an alternative leadership status as an umpire willing to call foul.

Kim uses the Security Council metaphorically as a sword China can use to attack, rather than as a shield to defend Third World interests. This is intended to illustrate what appears to be confusing, nonconforming behavior in the UN, but fits well with a behavior and paternalistic based definition of Chinese power. If China is attacking, is not that as much a leadership act in defence of others, as is more obvious shielding, protective forms? If China should be using a shield as Kim suggests, then there must be an attack to defend from, and all military strategists know that the best defense is a good offense. To attack relationships is to consider the human, behavioral, and power-based nature of politics and to focus on an alternative leadership source - China.

Lastly, Kim believes that China's passive Security Council behavior and pro Third World rhetoric seem to be contradictory and an attempt to placate both superpower and Third World interests and thereby is not a strong individual leader. We must remember however that the individual is defined in terms of relationships with others. Within the Security Council China does voice its opinions at the times and places that it feels appropriate, but it will not become a disruptive or coercive force out of deference to the principles embodied by the

UN charter and its sense of modeling an alternative and exemplary form of leadership within the UN. In Kim's example of China's actions and apparent conflict between reconciling principles and self-interest in the case of Angola, he looks for continuity where in fact each situation needs to be evaluated not for continuity, but for how China defines itself in relationship to the prevailing power relationships, its own abilities as power in those situations and how the problem is defined. Such a reexamination would involve a questioning of the analyst's conceptual framework for interpretation and a more full consideration of alternative definitions of power, leadership, and how the problem is defined from the Chinese political context. How are the rules and problems being interpreted tactically for strategic gain?

China not only opposed Soviet actions in Angola but also acknowledged the OAU request to remove military advisors and supported more broadly shared responsibilities for postwar compensation. In this way China worked within the existing set of power relationships to meet their political goals of antihegemonism by the USSR and to support Third World interests as far as possible within the larger group decisions of the UN as a nondisruptive member. Rather than labeling this as a 'have your cake and eat it too' position that implies incongruity and weakness, it reflects a more sophisticated and independent form of interaction by China within the UN. By looking for continuity in what the Chinese say and do Kim also leaves open the question of whether or not there is continuity in what the Americans say and do; how do they reconcile principles and interests; can they be separate? The question should be why do Americans perceive incommensurability between Chinese interests and principles rather than to summarily conclude that the Chinese are duplicitous.

As a paternalistic leader China is neither passive nor contradictory. For Chinese leadership within the context of the larger authority and responsibility of the UN, passive voting action is the only behavior option which reflects dissatisfaction, without coercion (a negative attribute) and yet accepts the decisions of the larger group. It is also not a contradiction between support for the Third World and actual Security Council resolutions. China's style and method of leadership are an acknowledgement of the Third World plight; while maintaining its own image and role as a superpower, China is an alternate leader to be emulated, above the hegemonic, coercive style political leadership employed by the US and former Soviet Union.

Jonathan Pollack makes the following observation:

> China has often acted in defiance of the preference or demands of both superpowers; at other times it has behaved far differently from what others expect. Despite its seeming vulnerability, China has not proven pliant and yielding toward either Moscow or Washington. Indeed in a certain sense China must be judged as a candidate superpower in its own right - not in imitation or emulation of either the Soviet Union or the United States, but as a reflection of Peking's unique position in global politics.[48]

Using the model-following concept this analysis has attempted to show that China considers itself to be a leader and an alternative to superpower pressure politics. China as a superpower and leader in the Security Council has acted as a supportive actor within the supranational authority of the United Nations, and as a self-appointed role model for other nations. China can be seen to be exercising strong leadership by abstention from voting in cases where it disagrees and yet is not disrupting the flow and necessary authority of the Security Council. There is deference to the larger group, but there is also an expression of opposition and individualism as an alternative style within the group. To see China as trying to define an autonomous leadership position is different qualitatively from the conclusion that it is cooperating by not interfering. To move past an interpretive framework that focuses on following rules as a definition of cooperation to one of how China may see itself in relationship to those rules changes the assumptions one must make in interpreting what is observed.

---

48. Jonathan Pollack, quoted in Paul Kennedy, *The Rise and Fall of Great Powers* (New York: Random House, 1987), 457.

From this perspective of China's relationship with the UN Security Council, China is not seen as a generally passive, sometimes contradictory actor who follows the rules and therefore is generally conforming, or as Jacobson and Oksenberg conclude that China will not 'subvert' these international institutions. Instead China is as assertive of its own position as the United States and Russia are of theirs, albeit through different means and definitions. China may be seen as cooperative in that it does not obstruct UN activities or violate accepted rules and norms of participation, yet cooperation also has the connotation of working together willingly for a common purpose. Given China's alternative leadership stance, it is not clear that China feels it is 'cooperating' simply because it has not violated any UN rules or obstructed the action of others.

Using the concept of model-following emphasizes questions about China's participation in the UN which Kim's analysis understated or overlooked. By seeing China from the perspective of model-following, its actions in the Security Council are more intentional and consistent, rather than a contradiction between principles and interests as Kim's analysis would lead us to believe. A linguistic analysis should complement our observations of political behavior; it forces one to take Kim's observations of China's behavior and his conclusion that it is largely symbolic and ask what it means and why it is meaningful rather than to dismiss it as unsubstantial. Further, if the Chinese are seen to behave symbolically, one should ask to why the symbolism has meaning, what it means and for whom, and how does that compare with symbolic American actions.

Some of the questions which this perspective leads me to ask are as follows: is China's compliance with the UN rules and procedures a conformity which overlooks or supersedes other Chinese perceptions of the UN? Exactly what meaning do the Chinese give to the notion of international cooperation as embodied in the UN, is it a forum which can perpetuate and enhance their legitimacy as an emerging world leader or is it only a source of domestic development programs? What will the relationship between a stronger, more stable domestic economic and political situation mean for China's self-perception as a model and therefore as a leader in international organizations? These questions highlight the point that compliance with the rules in not necessarily a cooperative act, but is perhaps one of immediate necessity, tactical actions for strategic gain. This fits with the idea of problem solving in the model-following concept were achieving a goal is primary, and each problem is dealt with as it emerges and is defined and resolved in terms of its relationship to achieving larger goals. This line of questioning should push one to examine the meanings of interaction with the international process more closely.

The critical common point between the model-following concept and neorealist definitions of cooperation is one of defining the rules, of defining what is acceptable, conforming behavior and, by exclusion, what is nonconforming. The critical difference is that when one only consider one's own culturally based definition of socio-political concepts as being acceptable to interpret all other experience within the international arena, one overlooks interpretations that not only provide different meanings but can lead to actions and changes in the international system.

Using the concept of model-following behavior brings up some interesting comparisons between neorealist conceptions of the international system and a model-based one. For both concepts the system and model-based social relations are made up of individual people represented by states that exist in relationship to one another. The individual is defined by the nature of the relationships, the basis of which is a function of power. Power is contained in knowledge and the ability to educate/socialize, to lead by example (Waltz says that weaker states will be motivated to emulate successful states). Education is a form of socialization and can also be the product of threats, punishment, promises and rewards; the Chinese are not naive about the uses of power.[49] Power is about the ability to create or

---

49. "China has very shrewdly and even brazenly used its available political, economic, and military resources. Towards the superpowers, Peking's overall strategy has at various times comprised confrontation, and armed conflict, partial accommodation, informal alignment, and a

enforce rules, either through overt actions or by creating laws, or both. The recent Chinese efforts to use military force to coerce and threaten Taiwan not to pursue independence is an example of their willingness to support their interpretation of the rules of sovereignty, and the American response was based upon different, alternative definitions of sovereignty and self-determination as a right of the Taiwanese. The difference lies in the interpretation of the rules and of power and cooperation that is necessary to achieve political goals.

Kim notes this difference when he asks if China "would deviate from the value-oriented model of world order and move in the direction of a power-oriented one?"[50] These concepts of values and power can not be separated as two different exclusive ideas. The problem that Kim fails to note is that American and Chinese interpretations of the international structure are both at one and the same time about values and power; the question is one of defining and interpreting what these concepts mean, for whom, and when. It is this arbitrary separation of values and power as two possibilities that Waltz makes in his defining the international system by virtue of states capabilities and a separation that prevailing definition of science makes by defining power as rational and values as irrational. The question should not be will China move from a value to a power-oriented model of world order, but how do their values and American values provide different interpretations of power embedded in the rules of the international structure. Power is embedded in language, and language as the medium through which interaction takes place is interpreted. To only focus on power as military or economic capabilities and potential it to ignore the interpreted meanings of power that occur daily and underlie statistical summaries of military or economic power.

---

detachment bordering on disengagement, sometimes interposed with strident, angry rhetoric. As a result China becomes all things to all nations, with many left uncertain and even anxious about its long-term intentions and directions." Jonathan Pollack in Paul Kennedy, *The Rise and Fall of Great Powers*, 457.

50. Kim, *China, The United Nations and World Order*, 501.

How does this idea of defining, interpreting, and following rules apply to Waltz's example of the shopkeeper who decides to stay open on Sundays? For Waltz the positive leader changed the rules by deciding to stay open on Sundays, creating an advantage in earned income and increasing its capabilities. The leader became stronger and the other shop owners had to respond, preferably by emulating the strong leader. Changing the rules, deciding to stay open on Sundays, changes the system. The process changed the rules and therefore the system which in turn affects the process, meaning that everyone has to interpret and react to the change. If, using the model-following concept, China's leadership, as the weaker store owners, realizes that they are not in a position to change the rules due to the increased strength of the strong leader to enforce them the leader becomes a model not only of how to increase one's capabilities but also of how to change the rules.

As the leader, the innovative shop owner expresses power through the formation of goals and rules/policy for achieving those goals: deciding to stay open on Sundays. Power for those following a model, the Chinese, lies in their interpretation of the rules and their resulting action both of which are based upon an understanding of themselves as a function of existing power relationships. The rules can be followed, openly disobeyed, or passively dealt with as a means of protesting and as a means of testing to see what can be gotten away with. In following the rules of the strong leader as a model the other store owners can exert power in several ways: (1) they can manipulate policy either through a deliberate misconstruing of the policy or by exploiting weaknesses in the new rules; (2) they breaking the rules as a means of experimenting and testing new policies; (3) they can achieve necessary production by using policy alternatives banned by the leader; and (4) they can use passive means such as sabotage or deliberate lethargy. On the other hand, the recognized governing authority does maintain ultimate power because there is a general deferment to the larger authority of the group, a recognition of existing social relations and power that can not be immediately changed. The effectiveness of these exertions of power by those following a model are not an overall change in the policy itself but can be a qualitative change in the degree to which the policy is effective or encompassing. The other store owners may not like now having to stay open on Sundays, but they can challenge that rule or rules in the present while working within the rules to increase their own power to alter the rule(s) more fundamentally at a future time.

Through the concept of model-following one might see the period of Maoism as a time of trying unsuccessfully to change the rules, and the current regime as following the model of the strong leader, the United States, to learn how to increase one's capabilities within the existing structure of rules to become strong enough to create and enforce a new set of rules.

Following rules while maintaining different beliefs is a culturally acceptable norm of interaction for the Chinese, whereas Americans tend to demand shared beliefs. A Western emphasis on shared beliefs encourages different measures of conforming than does a Chinese emphasis on correct behavior. Americans, as illustrated in the case studies I have presented, tend to see the Chinese as following the rules and therefore, as a measure of socialization to international, neoliberal norms, cooperating through acceptance of negotiated policy coordination. The Chinese however see themselves as following the principles embodied in the rules and therefore as a strong leader that exhibits correct behavior in the spirit of international cooperation as they define it. They are interpreting the rules to fit with their principles of leadership and power just as Americans use the rules to support our own interpretation. For the Chinese this is not cooperation as a socialization to specific Western liberal norms and values, as Waltz and Keohane believe that cooperation as they define it leads to. To believe that it is to maintain too narrow a definition of power as it is embodied in the prevailing analytical concept of 'cooperation.' The Chinese are interacting with the international system. They are actively involved to achieve the status and power of a developed state. This sounds 'realist' enough, but to stop here is to fail to consider other cultural definitions and concepts about the world of international politics.

Samuel Huntington in his article "The Clash of Civilizations?" states that "the great divisions among humankind and the dominating source of conflict will be cultural."[51] Huntington is critical of current prevailing attitudes in the West that illustrate and perpetuate the cultural divisions in the world today; it is a criticism that echoes the themes set out in this book.

> In the contemporary world most modern societies have been Western societies. But modernization does not equal Westernization. Japan, Singapore and Saudi Arabia are modern, prosperous societies but they clearly are non-Western. The presumption of Westerners that other peoples who modernize must become 'like us' is a bit of Western arrogance that in itself illustrates the clash of civilizations.[52]

He goes on to say that

> the West in effect is using international institutions, military power and economic resources to run the world in ways that will maintain Western predominance, protect Western interests and promote Western political and economic values.[53]

It is from this predominantly Western hold on global economic and military power that he sees two sources of conflict emerging, one is in the struggle for military, economic and institutional power, and the other more basic source of conflict will be in the differences between cultural values.

> Western concepts differ fundamentally from those prevalent in other civilizations. Western ideas of individualism, liberalism, constitutionalism, human rights, equality, liberty, the rule of law,

---

51. Samuel P. Huntington, "The Clash of Civilizations?" *Foreign Affairs* 72, no. 3 (Summer 1993), 22.

52. Samuel P. Huntington, "If Not Civilizations, What?" *Foreign Affairs* 72, no. 5 (Nov/Dec. 1993), 192.

53. "Clash of Civilizations?" 40.

democracy, free markets, the separation of church and state, often have little resonance in Islamic, Confucian, Japanese, Hindu, Buddhist, or Orthodox cultures. Western efforts to propagate such ideas produce instead a reaction against 'human rights imperialism' and a reaffirmation of indigenous values....The very notion that there could be a 'universal civilization' is a Western idea, directly at odds with the particularism of most Asian societies and their emphasis on what distinguishes one people from another. Indeed 'the values that are most important in the West are least important worldwide.'[54]

Huntington concludes his article with two policy prescriptions. Firstly, policies for short-term advantage that are based upon the need for Western civilization to "maintain the economic and military power necessary to protect its interests in relation to these civilizations" (non-Western civilizations, namely, Islam and Confucianism), and second, policies of what he calls accommodation. Accommodation means that the West must "develop a more profound understanding of the basic religious and philosophical assumptions underlying other civilizations and the ways in which people in those civilizations see their interests. It will require an effort to identify elements of commonality between Western and other civilizations."[55]

With supporting sentiment Joseph Nye, Jr. comments that the problem for the United States in the coming decades is not a challenge to American power, but "that the United States will have to adapt to new patterns of interdependence and new political agendas in the 21st century."[56] Unless we move beyond privileging our own definitions and include them as one of many possible cultural definitions, we will not be able to adapt. To paraphrase John Lennon, politics is what

---

54. Ibid., 40-41. Huntington cites Harry C. Triandis, *The New York Times*, 25 Dec. 1990, 41, and "Cross-Cultural Studies of Individualism and Collectivism," Nebraska Symposium on Motivation, vol. 37, 1989, 41-133.

55. Ibid., 49.

56. Joseph Nye, Jr., *Bound to Lead: The Changing Nature of American Power* (New York: Basic Books, 1990), 170.

happens while you're making other plans. To see Chinese international interaction from the perspective of model-following is one way to open up a new range of interpretations about what we observe that allows for a broader, more sophisticated understanding of their involvement in, and interpretation of the international arena.

China's behavior rather than being passive and at times contradictory is from the perspective I have described an active, logical leadership that sees itself as establishing an alternative model for international leadership by virtue of its independent courses of policy action. The Chinese are learning from the spirit of the American model of leadership and development, and following from this point of view believe in their ability to surpass the American model and become a preeminent model in their own right. As such they will be in a position to redefine the nature of relationships and thereby capable of reformulating goals and rules; exercising power in a new manner.[57] A goal of achieving the status and power of a developed nation has been established, a model selected to follow and learn from, and problems are defined and resolved in terms of the degree to which they inhibit achieving the goal.

The Chinese, like other non-Western nations, are doing what they believe is necessary to increase and improve their military and economic power. In doing so they are asserting their right for national security and economic development; rights they feel are being selfishly constrained by the United States. With regard to China/U.S. relations we can see this self-assertion of the right to develop economic and military security in China's sales of nuclear weapons to Pakistan, arms sales to the United States, conflicting interpretations over human rights, Chinese assertions of power in what they believe to be domestic issues in Taiwan and Tibet, and a growing willingness to not be intimidated by American threats of

---

57. As Paul Kennedy notes, "the more that China pushes forward with its economic expansion, the more that development will have power-political implications....It is only a matter of time." Paul Kennedy, *The Rise and Fall of Great Powers*, 458.

232

economic retaliation by increasing economic cooperation with other countries such as France.

I will briefly outline some of the key themes to emerge from the increased tensions between China and the United States as a means of illustrating the types of questions my concept of model-following encourages me ask and why I believe they are important questions. To move into a detailed policy analysis is not my point here, that is another book. I do however believe that it is from work in comparative policy analysis that the generalizations for future theories of international relations will emerge.

"China: Friend of Foe?"[58] For me, this bold headline summarizes American uncertainty about our relationship with China, concerns that were brought to the surface by China's barricade of Taiwan in an attempt to influence the March 1996 presidential elections on that island. The paradox for Americans is that China is being labeled as the new economic superpower of the next century, and yet it is a country that many perceive to not be following the current rules and norms for international behavior.[59] Gerald Segal of London's International Institute of Strategic Studies is quoted as saying that "China is even scarier than Japan" because at least Japan shares Washington's geopolitical goals.[60] An unnamed senior U.S. official is concerned because current U.S. and European trade policies towards China are serving to create "a competitor of extreme proportions, but we don't have the tools to deal with that threat."[61] As a last example, Marcus Noland of Washington's Institute of International

---

58. *Newsweek*, 1 April 1996, 24.

59. "Rethinking China," *Business Week*, 4 March 1996, 59.

60. Ibid., 64.

61. Ibid.

Economics says that because of China's enormous size and economic potential "it is the single biggest potential threat to the world trade system."[62]

This fear of the unknown China is complemented by confusion about China's actions. The following quote by U.S. Defense Secretary William J. Perry expresses the confusion many Americans feel towards China and the fear of a nation that is so unknown and that acts so independently. "Our policy accepts China at its word when it says that it wants to become a responsible world power, but China sends quite the opposite message when it conducts missile tests and large military maneuvers off Taiwan, when it exports nuclear-weapons technology or abuses of human rights."[63] With this statement Americans are being cast as open and trusting and the Chinese are perceived to be violating this openness. The problem with this statement is in not questioning what exactly a responsible world power means. Do we sell arms more responsibly than the Chinese? China may well want to be a responsible world power, however China's leadership does not view the American model of world leadership as being one of responsibility. So from a Chinese view being responsible is to *not* be like the Americans, which means adhering to Chinese goals and principles and finding means that work for them in the face of U.S. opposition. Unfortunately what the Secretary of Defense implies by saying that he takes China at their word, through his interpretation of what is 'responsible,' is that he assumes the Chinese *should* come to share our interpretation of what is responsible and become more like us in following the rules and goals of international behavior as we define and enforce them.

China's demonstration of force towards Taiwan has motivated American and Japanese leaders to focus on the issue of "how to handle the behavior of the

---

62. Ibid.

63. Ibid., 57.

Chinese."[64] It is China's threat of force and increasing strength to exercise force that most occupied the pages of the press, and often this increased strength was dismissed as the excessive nationalism that has replace Marxism as the leading state ideology. In addition to a general fear of China's military and economic potential, the themes I found in the American press reveal the degree to which the root of American political understanding is grounded primarily in military and economic factors of power and is an expression of our own nationalistic concerns for security on our terms.

China is a growing military and economic power that we as Americans perceive as an increasingly real threat to the current world order, this is not only the popular interpretation of the Chinese blockade of Taiwan and other recent acts, but it is also the neorealist academic interpretation. My point is not that China is innocent, their leaders are dissatisfied with the American dominated world system as it now exists and with continual American involvement in Chinese domestic affairs. However I would also propose that America's continued policy focus on China as a threat, and our fear of that threat is not a productive basis for generating policy action as it only serves to further entrench existing perceptions and antagonisms.

China is a power that many observers perceive to be operating outside of existing international norms and so is a threat to the rules as Western culture defines and enforces them. The present answer to this dilemma is to meet the threat with increased force, to 'handle the behavior of the Chinese' in an attempt to socialize them to what we as Americans consider to be proper behavior. This is the hope of increased economic ties with the Chinese and their involvement in international organizations, that somehow once they are interdependent then they will be subject to following the rules and punishments for not following the rules. I believe that for the United States to continue on this path will be increasingly

---

64. "Asians Feel Uneasy as China Rattles the Neighborhood," *Christian Science Monitor*, 13 March 1996, 6.

frustrating and tensions will reach unacceptable proportions, that is to the point of conflict. Already, China ignores U.S. threats of sanctions and by making decisions such as awarding France's Airbus Industrie worth U.S. $1.89 billion for 33 aircraft as a political snub is demonstrating an independence of action that is increasingly disconcerting for Americans.[65]

How does the concept of model-following help? There are four major subsets in this concept, (differential social relations, didactic social power, problem solving, and correct behavior) and I will briefly look at each one in as a potential source for generating new forms of dialogue with the Chinese. Model-following is first and foremost based upon the differential nature of social relations, where individuals are defined by the relationships that they are a part of. It is from within those sets of relationships that one can define goals of what to become and then selecting people already in those positions as models to follow and improve upon. How the Chinese perceive themselves in relationship to the United States and how we perceive them in relationship to us defines in large part their international identity and subsequent behavior. As long as the Chinese are treated as politically unequal (a regional power, and as seen earlier in this chapter as a weak international leader), and viewed primarily as an economic gold mine for the American economy the Chinese will shape their actions accordingly. They will continue to struggle to show that they are not what we define them to be. The goal for the Chinese is to be a strong international power both economically and militarily, and to no longer be ignored as they historically have been. In this way we might think of the Chinese as working within an interpretation of international norms that supports their goals, much as American leaders do to define and achieve goals.

We as Americans need to fundamentally alter the basic assumptions of our relationship if we are to move past fear as the basis for our policy making. The

---

65. "In Slap at U.S., China Rewards France for Downplaying Rights," *Christian Science Monitor*, 12 April 1996, 6.

question that this leads me to ask is: what points of national identity and policy goals have commonality and difference as expressed in Sino/American policy dialoge and how can the linguistic examination of the dialogue as a symbolic act lead us to generate new metaphors with more closely shared meanings that can alter the basis of our relationship beyond those of economics and security as they are now expressed?   This question can only be dealt with by first seeing the Chinese as making one interpretation of existing international rules and norms, and Americans as making another. To see the Chinese as outside is to set up the false dichotomy of they are wrong and we are right, and to overlook China's active, consistent involvement in a variety of international organizations and treaties. Examining how these institutions support alternative identities and interpretations will also lead to seeing points of commonality that will be the basis for alternative, future interpretations.

The second component of model-following is power. Power is grounded in the diadactic nature of social relations, that is differential relationships based upon teaching and learning. The U.S. is now in the leadership position and is viewed by China as a model to learn from. This does not mean copying, but learning that which is beneficial, and learning from the mistakes inherent in the model as points to improve upon so as to become in the long term better than the model; a model in one's own right. The Chinese are working assiduously towards their goal of modernization as they have defined it, American observers would do well to ask what is being absorbed and what is being tossed aside, and how is Chinese culture altering that which is taken in? What does it mean for them? This line of questioning should move past technology and development plans, and consider more fully concepts from Western culture and politics that will shape Chinese political dialoge and society, but as this book has taken great pains to point out American observers must be willing to let go of their own definitions of these concepts for the sake of analysis. We can not look for indicators of liberalization, or democracy without first asking what these concepts mean for the

Chinese. To do otherwise is to continue along the present path of trying to validate Western definitions of these concepts. I also would ask what are Americans, as the 'model' of a world leader, teaching through their actions to the Chinese, what do we want to teach them? This also includes the lessons that the Chinese learn by watching U.S. actions in a variety of global and domestic situations.

To be interested in trying to learn what it is the Chinese want to achieve for themselves as they define it would also be a way to acknowledge that they may come up with a unique and culturally satisfactory means of governing, that they may be a model for Americans to learn from. Considering that the Chinese are currently battling the issues of overpopulation and environmental/resource degradation on a scale that we have yet to face this would not be unreasonable. To do this would send a message to the Chinese that we view them in a different way, not as a secondary power, but as a cultural equal, and one worth learning from. The Chinese whom I spoke with took no small measure of pride in every instance where other people's had come to learn from them. To acknowledge the Chinese in this way would not diminish American power, but would enhance America's position as a model and as a leader. Recall Mao's advice to learn from the people in order to be a better teacher.[56]

This leads to the third point, problem solving. Every policy and every action is a part of the Chinese experiment in how to quickly and effectively achieve their goals of development. For example, selling arms is a quick way to earn much needed hard currency, and given that the U.S. is the world's largest arms dealer it could be said that they are following our lead. We protest their activities because their arms sales are to potential enemies, but on the other hand

---

66. This recalls the admonishon of Dr. James C. Yen, a Chinese reformer in the 1920's who started an education revolution in China, and his philosopy is alive today through the work of the Philipean Rural Reconstruction Movement, and the International Institute for Rural Reconstruction. His successful work and philosophy are summarized in the work by Dr. James B. Mayfield, a student of Dr. Yen's in the book, *Go to the People: Releasing the Rural Poor through the People's School System* (West Hartford, Conn: Kumarian Press, 1985).

the Chinese have as creative entrepreneurs and smart capitalists found their market niche. Americans oppose and protest Chinese actions through a variety of means and the Chinese find new ways of achieving the same ends. I believe that these protests are Band-Aids, only raising new details to be enforced within existing rules rather than addressing issues of goals and identity behind the specific actions. It is the emphasis that we put in the rules through our interpretations and enforcement of them that push the Chinese to new levels of creativity in problem solving towards the same goal. When we threaten the Chinese with sanctions, or harsh diplomatic language they do not see U.S. policy language as the behavior modifying tool we intend it to be. Instead their view of the U.S. as an imperialist power, and a poor model of an international leader is reinforced.

Recall the struggles of the Chinese government in dealing with the peasants over grain prices as I outlined in Chapter 3, it was a constant effort of outmaneuvering by the peasants and preventive action by the government. The issues of human rights, copy rights, nuclear proliferation, arms sales, these are no different. They are experiments, problem solving in order to gain technology and earn hard currency all towards the goal of development, some will work better than others and some will not work at all. The Chinese are trying to find the path through the mountains, trusting that it will emerge only as they proceed towards their goal. American efforts to 'manage Chinese behavior' only makes the path more difficult, but does not alter it or the goals.

As the Chinese government learned in their dealings with the peasants the tolerance for overt pressure through coercion has its limits, it was through creative policy making that the government maintained its position. In American dealings with China overt pressure will also have its limits, the Chinese can not be coddled but we must develop a basis for mutual respect. Again, this must be based upon changing our dialogue with them about the fundamental way that China and the United States view themselves, one another and various world problems. This is more than work for diplomats at the UN or in the State Department who only maintain and perpetuate current metaphors in our dialogue with China. Looking at U.S./China relations requires careful analysis of language as the carrier of culture and meaning that can give the politicians of both nations a new language and different perspectives to use.

Lastly there is the issue of correct behavior, of maintaining 'correct' outward appearances based upon principles. If Taiwan does not officially declare independence correct behavior in terms of the proper relationship between China and Taiwan are maintained, and the continued possibility for diplomatic resolution based upon the current principle that there is only one China is maintained. Perhaps a new model for their relationship can be developed from the metaphor of 'like a family' that I found to be important in Chapter 4, and how if at all to maintain one family with a rebel child? What metaphors work with the Chinese, are they interested in seeing a family counselor to work out their disputes? The Philippines government found diplomatic channels to be effective in dealing with the Chinese in the Spratly Island territorial dispute, they appealed to the Chinese sense of wanting to follow international law in a multilateral forum.[67] Recall the earlier discussion of the Chinese preference for open forums rather than backroom deal making as they see the U.S. doing. 'Correct behavior' is a form of ritual activity, and it is the product of the differential structure and didactic nature of social relationships. The U.S. and China need to come to terms with new sets of 'correct behaviors' that break the current pattern of interaction and which can only be base upon new relational definitions and a willingness by both societies to adapt to a newly emerging international situation. Finding acceptable rituals will again be the product of increased cultural awareness by both parties.

---

67. "Manila Treads on Eggshells with China Over Ship Run-in," *Christian Science Monitor*, 2 February 1996, 7. "The Philippeans' diplomatic offensive to get international opinion on its side after discovering the Chinese occupation of Mischief Reef has paid off. China has since agreed to discuss the Spratlys claims with all claimants within the context of the United Nations Law of the Sea and in a multilateral basis."

The questions I have raised are important because they move past power based forms of analysis. Power based analysis is important for issues of security and defining threats, but it is based largely on fear and perpetuates an ethnocentric world view that prevents or greatly slows the process of understanding, or accommodation to use Huntington's term, which is equally important for adapting to new global environments. To only consider the power based perspective is to believe that China challenges the existing rules of international behavior and Western nations need to respond in kind or with greater force. Although military aggression cannot be tolerated, to maintain only a power based view is to believe naively that through force and threat of force whether economic, military or diplomatic that Western policy makers can change or 'manage' Chinese behavior and their strategic goals. Using culturally based concepts to help us understand possible Chinese interpretations of the world is to move beyond the exclusive consideration of power as Americans now define it. There is a time and a place for the threat of and uses of force, but to rely too much on these rough tools is to ignore yet untapped potentials for interaction.

I believe that it will be through the careful analysis of current Chinese/American policy discourse as the carrier of cultural values, and historical influences, the understanding of which is mediated by interpretation, that new perspectives and fresh dialogue can begin. To focus on power is to only focus on the status quo and threats to the existing order as the existing 'strong powers' define it and its enforcement. To include cultural issues of meaning is to open up the possibility for stimulating new dialogue that moves past the negative and downward cycle of potential threat and the fear of retaliation which leads to real threats and real retaliation as both parties become increasingly frustrated at the other's intransigence.

To begin a new dialogue with China must mean treating them as the equal that our deep seated fear implies them to be; fear that is based upon a general American confusion about who the Chinese are as a people and as an emerging world power. We could, for example, ask the Chinese what they mean by a responsible world power, which they have already articulated, open our own definitions up for discussion and begin a dialogue towards a new form of relationship. This does not mean sacrificing our values or goals, but it does mean that we have to be as willing to adapt to a changing world as we expect the Chinese to be; a world that is being defined by the present nature of relationships and the patterns of discourse in the international arena.

Conclusions

This book has established several important themes that are active in the implementation of Ecological Agriculture as a national policy. It is a policy that is symbolic of China's current development dilemma, and one that is symbolic of where many would like to see China go in the future. Through an interpretive analysis I have tried to show what meanings this policy has for the Chinese involved and how the cultural themes that emerged challenge American analytical concepts. As social scientists we should not be looking for signs of democracyy, or capitalism, or cooperation all of which are founded on our own definitions and the assumption that these developments will hasten China's interdependence and diminish it as a political threat. I believe that the more important and in the long term more significant forms of analysis will provide us with insights into what cultural and political meanings these concepts have for the Chinese and how this will increase our understanding of contemporary Chinese society and of our own American culture. The fall of Communism in China will not bring us closer to understanding Chinese politics, our concepts will still be inadequate. Broadening the interpretations that our analytical concepts provide us by incorporating meaning from the perspective of those we observe must be considered.

The model of Communism as envisioned by Mao has been deemed by Americans and Chinese alike as unsuccessful and self-defeating for achieving social and political development. The challenge of the CCP today is to quickly redefine and clarify its goals, establishing itself as a model for the twenty-first century.   Should they fail a new government will just as quickly have to show itself to be a viable model, that is it must establish its authority to define goals and generate models to follow while continuing to maintain social order and unity. What we are failing to apprehend are the meanings and the rituals that will have significance for the Chinese and a new government. Our lack of appreciation comes in the form of seeing the Chinese actions and interpretations as a threat or nonthreat to our own power, without considering other possibilities for categorization, and dialogue.

How can we acknowledge different interpretations of the concepts of power, of cooperation, and of leadership that are based upon the different historical experiences and cultural identities of the Americans and Chinese. The assumptions embedded in our concepts is that if the Chinese follow the existing rules of the international structure they will come to share our interpretation of them, becoming cooperative partners rather than a threat to the system as we define it. The problem that is posed with the concept of model-following is that what we perceive as cooperation is for the Chinese a means to an end; a goal that potentially conflicts with the current status quo and an American interpretation of the ideal goal. It can be seen as the Waltzian goal of survival and emulation of strong states to become a strong state. It is a potential conflict because of the challenge it makes to an America as a strong state and the interpretations that we give to the rules we make to ensure our survival. The neorealist use of the concept of cooperation rests upon the belief that the process of interaction, of friendly cooperation by creating rules to follow, will lead to shared values. By following the 'right' process mutually satisfactory goals will be achieved.

This book also shares this liberal assumption by proposing that it is through conceptual self-reflection, changing the nature of our assumptions and thereby improving understanding and dialogue that through the process of communication conflict can be minimized so that mutually satisfactory goals can be achieved. Rather than defining states solely in terms of power capabilities, we as analysts need to open up, or unpack our own Western political and economic concepts that define goals and processes for evaluation and consideration as one of many possible interpretations. This is not to dismiss our ideals and concepts as useless or irrelevant, but to question how they can become more expansive and inclusive, useful for understanding, analysis, and stimulating dialogue that is more inclusive rather than tending towards only defining potential enemies. Americans and Chinese say they want international cooperation and an equitable international economic system, but the interpretations given to these goals and processes as concepts are different. By understanding these different meanings and the metaphors and rituals used to give them meaning, new concepts can emerge to facilitate dialogue and debate. The concept of model-following defines different interpretations of power, leadership and cooperation that need to be negotiated with our own interpretations of these concepts to stimulate new thinking and dialogue. As Keohane puts it, we need to move past "dogmatic assertions of epistemological or ontological superiority" to find "ways of discovering new facts and developing insightful interpretations of international institutions."[68]

I think that if we continue to use Western liberalism as a standard of measure for democratization in China that the measure of and metaphors used need to be adjusted to the realities of Western liberalism as we move into the twenty-first century. If the Chinese see America as a model of an advanced country, how do they see the model, its inherent contradictions, and the possibilities to improve upon it? An understanding of Chinese politics would be well complemented by a more thorough understanding of ourselves as we are moving through our own post-1989 social, cultural, political and economic crises. If our observations of increasing social and economic diversity in China are interpreted as emerging pluralism that is believed to imply a potential base for

---

68. Keohane, *International Institutions and State Power*, 174.

democracy, what does it mean for our democracy when increasing numbers of people feel excluded and are also demanding political expression? If China follows the rules of the international organizations of which it is a member and this is considered to be cooperation, what does it mean when we don't follow the rules and are not in good standing are we uncooperative, or taking a symbolic leadership position? If it is believed that China's ability to translate Third World identity into interpersonal cooperation is hampered by its racism, what does that say about the effect of American racism and our ability to identify and cooperate with other races and nations, including China?

These issues are not as far apart as they seem. It is the questions and the concepts that we use for analysis that keep us separate. "Both the distinctiveness and non-distinctiveness of cultures are not facts about cultures, but artifacts of the ways in which cultures are studied."[69]   The point is that if we are studying political and economic change in China, if we are looking to see how China 'matures' and becomes 'modern' and if we see ourselves at a different, more advanced 'stage' of development we will assume that China will become more like us. In these concepts there is the assumption that as the strong power the United States and the current rules of the international structure as we interpret them are the best model to be followed, just as the store keeper that initiated the change to stay open on Sundays. If we see ourselves as defining one possible avenue of development and maturity, and ask what other definitions we can find that are grounded in contemporary policy action, different observations, comparisons, and conclusions will emerge. To be value free in the social sciences ought to mean a recognition and freeing of one's own values to be thrown into the analytical pot along with everyone else's to find new concepts, metaphors and questions. The result is not the feared loss of cultural identity and political ability, but an increase in understanding and awareness.

---

69. Andrew J. Nathan and Tianjin Shi, "Cultural Requisites for Democracy in China," *Daedelus* 122, no.2 (Spring 1993), 118.

Politics is a symbolic venture, the rituals, metaphors, and myths that evolve define social groups, interpret what is observed and justify actions. If a global economic and political environment that is based on Western liberalism and market values is what we believe that we desire for ourselves and others those concepts will generate the metaphors we use for observation, comparison, interpretation, and action. How these concepts and others are defined by international actors shapes the interpretations we make through their use.

The future is understood through interpretations of the present, and rarely, if at all, ever meets one's expectations of what should happen. People were shocked when the Soviet Union fell apart, and they have been amazed at the turnarounds in American domestic politics. The shock and the amazement are because they could not predict what would happen; their analytical concepts did not serve their purpose. Rather than asking about other interpretations of meaning, one assumes the cultural validity and interpretation of these concepts and asks for predictions based upon an old model of science. Americans are waiting for the fall of the Chinese Communist Party because that is what happened in the Soviet Union and East Europe, and so now they are training their analytical gaze on the indicators of liberalism. Will we do better this time? No one ventures a guess. They might, however, try to understand what they are seeing more fully and through that understanding use their position as social scientists to stimulate new thinking and new discourse for the emerging international relations of the twenty-first century.

SELECTED BIBLIOGRAPHY

Addler, Emmanuel, ed. *Progress in Postwar International Relations*. New York: Columbia University Press, 1991.

Aggarwal, Vinod K. "The Unraveling of the Multi-Fiber Arrangement 1981: An Examination of International Regime Change." *International Organization* 37 (Spring 1983): 617.

Agrawal, Ramesh C. "Resource Use/Management for Sustainable Agriculture: Some Experiences in Maguyu Project Shandong Province, China." In *Integrated Resource Management For Sustainable Agriculture: Proceedings of the International Conference*, eds., Shi Yuanchun and Cheng Xu, 275-287. Beijing: Beijing AgriculturalUniversity Press, 1994.

Azizur Rahman Khan, Keith Griffin, Carl Riskin, Zhao Renwei. *Household Income and Its Distribution in China*. Working paper, Department of Economics, University of California Riverside. December 1991.

Ball, Terence. "Sinn and the Social Scientist: Deadly Hermeneutics." In *Idioms of Inquiry*, ed. Terence Ball, 108. Albany: State of New York Press, 1987.

Banister, Judith. "China's Population Changes and the Economy." In *China's Economic Dilemmas in the 1990's: The Problems of Reforms, Modernization, and Interdependence*, Joint Economic Committee, Congress of the United States, eds., 234-251. London: M.E. Sharpe, 1992.

Blaikei, Piers, and Harold Brookfield. *Land Degradation and Society*. New York: Routledge, 1987.

Bleicher, Josef. *Contemporary Hermeneutics: Hermeneutics as Method, Philosophy and Critique*. London: Routledge and   Kegen Paul, 1980.

Bochuan, He. *China on the Edge: The Crisis of Ecology and Development*. San Francisco: China Books and Periodicals Inc., 1991.

Bond, Michael, ed. *The Psychology of the Chinese People*. Oxford: Oxford University Press, 1986.

248

Borden, George A. *Cultural Orientation: An Approach to Understanding Intercultural Communication*. Englewood Cliffs, N.J.: Prentice Hall, 1991.

Brinkerhoff, Derick W. and Arthur A. Goldsmith. "Promoting the Sustainability of Development Institutions: A Framework for Strategy." *World Development Journal* 20   (1992): 369-384.

Buker, Eloise. *Politics Through a Looking Glass: Understanding Political Cultures Through Structuralist Interpretations of Narratives*. Westport: Greenwood Press, 1987.

Burch, Betty. "Models as Agents of Change in China." In *Value Change in Chinese Society*, eds., Richard W. Wilson and Amy Auerbacher, 122-137. New York: Praeger Press, 1979.

Cahn, Matthew Alan. *Environmental Deceptions: The Tension Between Liberalism and Environmental Policy Making in the United States*. Albany: SUNY Press, 1995.

Cao, Dexiang. *Public Administration in Post-Mao China (1978-1988): A Theoretical Analysis Through Rosenbloom's Model*. Ph.D. dissertation, State University of New York at Albany, 1990.

Carlsnaes, Walter. "The Agent-Structure Problem in Foreign Policy Analysis." *International Studies Quarterly* 36 (1992): 245-270.

______.*Ideology and Foreign Policy: Problems of Comparative Conceptualizations*. Oxford: T.J. Press Ltd., 1986.

Chan Wing-Tsit. "Chinese Theory and Practice." In *The Chinese Mind: Essentials of Chinese Philosophy and Culture*, ed., Charles A. Moore, 11-30. Honolulu: University of Hawaii Press, 1967.

Chay, Jongsuk, ed. *Culture and International Relations*. New York: Praeger, 1990.

Chen Jiefu, Dongye Guangliang, and Feng Yongjun. "Developing Sustainable Agriculture to Promote Agricultural Modernization." Unpublished paper, est. 1994. Shandong Agricultural University, People's Republic of China.

Cheng Xu. "'Hard' and 'Soft' Restraints for China to Sustain Agricultural Development and to Follow the Conventional Modernization Approach and Deserved Alternative Way." In *Integrated Resource Management For Sustainable*

Agriculture: Proceedings of the International Conference, eds., Shi Yuanchu and Cheng Xu, 412-417. Beijing: Beijing Agricultural University Press, 1994.

Chilcote, Ronald. "Theories of Political Culture: Collectivity and the New Person." In *Theories of Comparative Politics*. Boulder: Westview Press, 1981.

China's Agenda 21: White Paper on China's Population, Environment, and Development in the 21st Century. Adopted at the 16th Executive Meeting of the State Council of the People's Republic of China on 25 March 1994. Beijing: China Environmental Press, 1994.

China's Eco-Farming. Compiled by the Chinese National Environmental Protection Agency and the United Nations Environmental Program. Beijing: China Environmental Science Press, 1990.

Cohen, Anthony. *The Symbolic Construction of Community*. London: Tavistock Publications, 1985.

Cohen, Benjamin J. "The Political Economy of Trade." *International Organization* 44 (Spring 1990): 261.

Cowhey, Peter. "The International Telecommunications Regime: the Political Roots of Regimes for High Technology." *International Organizations* 44 (Spring 1990): 169.

Czempiel, Ernst-Otto and James N. Rosenau, eds. *Global Changes and Theoretical Challenges*. Lexington:Lexington Books, 1989.

Deng, Lang. *Internal Migration and the Development of China, 1982-1987*. Master's thesis, University of Utah, 1991.

Des Forges, Roger. "Democracy in Chinese History." In *Chinese Democracy and the Crisis of 1989: Chinese and American Reflections*, eds., Roger V. Des Forges, Luo   Ning, Wu Yen-bo, 21-52. Albany: SUNY, 1993.

Dittmer, Lowell, and Samuel S. Kim. *China's Quest for National Identity*. Ithaca: Cornell University Press, 1993.

Donnelly, Jack. "International Human Rights: A Regime Analysis." *International Organization* 40 (Summer 1986): 599.

Duara, Prasenjit. *Culture, Power, and the State: Rural North China 1900-1942*. Stanford: Stanford University Press, 1988.

250

Ebel, Roland H. and Raymond Taras. "Cultural Style and International Policy-making: The Latin American Tradition." In *Culture and International Relations*, ed.Chay, Jongsuk, 191-266. New York: Praeger, 1990.

Eckstein, Harry. "A Culturalist Theory of Change." *American Political Science Review* 82 (September 1988): 789-804.

Edelman, Murray. *Politics as Symbolic Action: Mass Arousal and Quiescence.* Chicago: Markham Pub. Co., 1971.

________. *The Symbolic Uses of Politics*. Urbana: University of Illinois Press, 1973.

Elder, Charles, and Roger Cobb. *The Political Uses of Symbols*. New York: Longman Press, 1983.

Esherick, J.W., and J.N. Wasserstrom. "Acting Out Democracy." In *Popular Protest and Political Culture in Modern China*, eds., Jeffery Wasserstrom and Elizabeth J. Perry, 28-66. Boulder: Westview Press, 1992.

Falk, Richard A. "Culture, Modernism, Postmodernism: A Challenge to International Relations." In *Culture and International Relations*, ed. Jongsuk Chay, 267-79. New York: Praeger, 1990.

Fei Xiaotong. *From the Soil: The Foundations of Chinese Society*. Translated by Gary G. Hamilton and Wang Zheng. Berkeley: University of California Press, 1992.

Fernandez, James W. "The Performance of Ritual Metaphors." in *The Social Use of Metaphor: Essays on the Anthropology of Rhetoric*, eds., David Sapir and J. Christopher Crocker, 95-123. Pittsburg: University of Pennsylvania Press, 1977.

Franz, M.L. von. "The Process of Individuation." In *Man and His Symbols*, ed., Carl Jung. New York: Dell Publishing, 1968.

Freeman, Barbara. "Implementation and Madisonian Government." in *Implementation and the Policy Process: Opening Up the Black Box*, eds., Dennis J. Palumbo and Donald J. Calista, 39-50. New York: Greenwood Press, 1990.

Frost, Peter J., Larry F. Moore, Meryl Louis, Craig Lundberg, Joanne Martin, eds. *Reframing Organizational Culture*. London: Sage Publications, 1991.

Fu, Zhengyuan. "Continuities of Chinese Political Tradition." *Studies in Comparative Communism* 24 (September 1991): 259-80.

Geertz, Clifford. *The Interpretations of Cultures*. Basic Books Inc., New York, 1973.

Gold, Thomas. "Urban Private Business and Social Change." In *Chinese Society on the Eve of Tiananmen: The Impact of Reform*, eds., Deborah Davis and Ezra Vogel, 168-181. Cambridge: Harvard University Press, 1990.

Gold, Thomas B. "After Commradship: Personal Relationships in China Since the Cultural Revolution." *China Quarterly* 12 (1985): 657-675.

Goldberg, Joseph R. "Grain Options for China: 1990-2000." In *Agricultural Reform in China*, T.C. Tso, ed., 114-122. Beltsville, Md.: Ideals Inc., 1990.

Goldman, Merle, Perry Link, and Su Wei. "China's Intellectuals in the Deng Era." In *China's Search for National Identity*, eds., Lowell Dittmer and Samuel Kim, 125-153. Ithaca: Cornell University Press, 1993.

Haas, Ernst B. "Words Can Hurt You, or Who Said What to Whom about Regimes." *International Organization* 36 (Spring 1982): 207-244.

Haas, Ernst S. "Regime Decay: Conflict Management and International Organizations." *International Organization* 37 (Spring 1983): 189.

Haas, Michael. "Asian Culture and International Relations." in *Culture and International Relations*, ed. Jongsuk Chay, 172-189. New York: Praeger, 1990.

Haas, Peter M. "Do Regimes Matter? Epistemic communities and Mediterranean Pollution Control." *International Organization* 43 (Summer 1989): 377.

Haggard, Stephan and Simmons, Beth A. "Theories of International Regimes." *International Organization* 41 (Summer 1987): 491.

Han Xiaoxing. "Democratic Transition in China." In *Chinese Democracy and the Crisis of 1989: Chinese and American Reflections*, eds., Roger V. Des Forges, Luo Ning, and Wu Yen-bo, 223-239. Albany: SUNY, 1993.

Harding, Harry. *China's Second Revolution: Reform After Mao*. Washington D.C.: Brookings Institute, 1987.

Harrison, Michael I. *Diagnosing Organizations: Methods, Models, and Processes*. London: Sage Publications, 1987.

Hawkesworth, M.E. *Theoretical Issues in Policy Analysis*. Albany: SUNY Press, 1988.

Hinton, William. *The Great Reversal: The Privatization of China, 1978-1989*. New York: Monthly Review Press, 1990.

Horowitz, Donald. "Cause and Consequence in Public Policy Theory: Ethnic policy and System Transformation in Malaysia." *Policy Sciences* 22 (1989): 249-287.

Huang, Philip C.C. *The Peasant Family and Rural Development in the Yangzi Delta: 1350-1988*. Stanford: Stanford University Press, 1990.

Huntington, Samuel. "The Clash of Civilizations?" *Foreign Affairs* 72, no. 3 (Summer 1993): 22-50.

______. "If Not Civilizations, What? Paradigms of the Post-Cold War World." *Foreign Affairs* 72, no. 5 (November/December 1993):186-194.

______. "The Problem of the Study of Political Change. "In *Comparative Political Systems*, ed. Louis   Cantori, 400-440. Boston: Holbrook Press, 1974.

Hwang Kwang-kuo. "Face and Favor: The Chinese Power Game. "*American Journal of Sociology* 92 (January 1987): 944-974.

Jacobson, Harold K., and Michel Oksenberg. *China's Participation in the IMF, the World Bank, and GATT*. Ann Arbor: University of Michigan Press, 1990.

Jervis, Robert. *Perception and Misperception in International Politics*. Princeton: Princeton University Press, 1976.

Ke Jianguo, Li Jincuo. "Chinese Sustainable Agriculture and Traditional Agriculture Systems." Unpublished Paper, est., 1994. Nanjing Agricultural University, People's Republic of China.

Kelliher, Daniel. *Peasant Power in China: The Era of Rural Reform 1978-1989*. New Haven: Yale University Press, 1992.

Kennedy, Paul. *The Rise and Fall of Great Powers*. New York: Random House, 1987.

Keohane, Robert O. *After Hegemony: Cooperation and Discord in the World Political Economy*. Princeton: Princeton University Press, 1984.

______. *International Institutions and State Power*. Boulder: Westview Press, 1989.

______. ed. *Neorealism and its Critics*. New York: Columbia   University Press, 1986.

Keohane, Robert, and Joseph S. Nye. "Power and Interdependence Revisited." *International Organization* 41 (Autumn 1987): 724-53.

Kim, Samuel. "Behavioral Dimensions of Chinese Multilateral Diplomacy." *The China Quarterly* 72 (December 1977): 713.

______. *China the United Nations, and World Order*. Princeton: Princeton University Press, 1979.

______. ed. *China and the World: New Directions in Chinese Foreign Relations*. Boulder: Westview Press, 1989.

King, F.H. *Farmers for Forty Centuries: Permanent Agriculture in China, Korea, and Japan*. Madison, Wis.: Mrs. F.H. King, 1911.

Kleinman, Sherryl, and Martha A. Copp. *Emotions and Fieldwork*. Qualitative Research Methods Series 28. London: Sage Publications, 1993.

Knudsen, Baard. "The Paramount Importance of Cultural Sources." *Cooperation and Conflict* 22 (1987): 81-113.

Krasner, Stephen D., ed. "International Regimes." *International Organization* 36 (Spring 1982).

Kratochwil, Friedrich and Ruggie John G. "International Organization: A State of the Art on an Art of the State." *International Organization* 40 (Fall 1986): 753.

Lakoff, George, and Mark Johnson. *Metaphors We Live By*. Chicago: University of Illinois Press, 1980.

Lampton, David, ed. *Policy Implementation in Post-Mao China*. Berkeley: University of California Press, 1987.

Lan Jusheng, Jia Rujiang, and Wei Jiankun. "The Concept and Strategies of Establishing Sustainable Agriculture in North China Lowland Plain." In *Integrated Resource Management For Sustainable Agriculture: Proceedings of the International Conference*, eds., Shi Yuanchun and Cheng Xu, 556-60. Beijing: Beijing Agricultural University Press, 1994.

Lavely, W.R. "The Rural Chinese Fertility Transition: A Report from Shifang Xian, Sichuan." *Population Studies* 38 (1984): 365-84.

Liddle, William. "The Politics of Development Policy." *World Development Journal* 20 (1992):793-808.

Lieberthal, Kenneth, and David Lampton. *Bureaucracy, Politics, and Decision Making in Post-Mao China.* Berkeley: University of California Press, 1992.

Lieberthal, Kenneth and Michel Oksenberg. *Policy Making in China: Leaders, Structures, and Processes.* Princeton: Princeton University Press, 1988.

Lin Zhiling. "Traditional, Modernist, and 'Party' Culture in Contemporary Chinese Society." In *The Chinese and Their Future*, eds., Zhiling Lin and Thomas W. Robinson, 134-149.

Lindenberg, Marc. "Making Economic Adjustment Work: The Politics of Policy Implementation." *Policy Sciences* 22 (1989): 359-394.

Link, Perry. *Evening Chats in Beijing: Probing China's Predicament.* New York: Norton Press, 1992.

Liu Shukai. "The Theme of Chinese Modern Ecological Agriculture." Unpublished paper, est. 1994. Nanjing Agricultural University, People's Republic of China.
Liu Sihua. "An Approach to the Problems concerning Environment Protection and Ecology Construction in Socialist Market Economy." *Sheng Tai Jing Ji* [Ecological Economics] 1 (February 1994): 1-14.

Love, Janice and Peter C. Sederberg. "Euphony and Cacophony in Policy Implementation: SCF and the Somali Refugee Problem." *Policy Studies Review* (Autumn 1987): 155-173.

MacIntyre, Alasdair. *Against the Self-Images of the Ages.* Notre Dame: University of Notre Dame Press, 1978.

Mao TseTung. "Preface to Rural Surveys." *Selected Readings from the Works of Mao TseTung.* Beijing: Foreign Language Press, 1971.

______. "Get Organized." *Selected Readings from the Works of Mao TseTung.* Beijing: Foreign Language Press, 1971.

______. "On Practice." *Selected Readings from the Works of Mao TseTung*. Beijing: Foreign Language Press, 1971.

______. "On Contradiction." *Selected Readings from the Works of Mao TseTung*. Beijing: Foreign Language Press, 1971.

Mayfield, James B. *The Egyptian Dilemma: The Challenge of Local Government in the Processes of Development, Decentralization and Democratization*. Chapter 1, unpublished manuscript, 1992.

______. *Go to the People: Releasing the Rural Poor Through the People's School System*. Hartford, Conn: Kumarian Press, 1985.

McLaughlin, Andrew. *Regarding Nature: Industrialism and Deep Ecology*. Albany: SUNY Press, 1995.

Morse, Janice M., ed. *Critical Issues in Qualitative Research Methods*. London: Sage Publications, 1994.

Mufson, Steven. "Why China Talks Tough and Flexes Its Military Muscle." *San Francisco Chronicle*, 25 March 1986, sec. A, 8.

Munro, Donald. *The Concept of Man in Early China*. Stanford: Stanford University Press, 1979.

Nadelmann, Ethan A. "Global Prohibition Regimes: the Evolution of Norms in International Society." *International Organization* 44 (Spring 1990): 479.

Nakamura, Robert T. "The Japan External Trade Organization and Import Promotion: A Case Study in the Implementation of Symbolic Policy Goals." In *Implementation and the Policy Process: Opening Up the Black Box*, eds., Dennis Palumbo and Donald Calista, 67-86. New York: Greenwood Press, 1990.

Nathan, Andrew, and Tianjin Shi. "Cultural Requisites for Democracy in China." *Daedalus* 122 (Spring 1993): 95-124.

Ng-Quinn, Michael. "National Identity and Premodern China." In *China's Search for National Identity*, eds., Lowell Dittmer and Samuel Kim, 32-61. Ithaca: Cornell University Press, 1994.

Nye, Joseph Jr. *Bound to Lead: The Changing Nature of American Power*. New York: Basic Books, 1990.

Olson, Alan M. *Myth Symbol and Reality*. Notre Dame: University of Notre Dame Press, 1980.

Ortony, Andrew, ed. *Metaphor and Thought*. Cambridge: Cambridge University Press, 1993.

Ott, Steven. *The Organizational Culture Perspective*. Pacific Grove, CA: Brooks/Cole Publishing, 1989.

Palumbo, Dennis J. and Donald Calista. *Implementation and the Policy Process: Opening Up the Black Box*. New York: Greenwood Press, 1990.

Park, Han-Sik. "Ideology and International Relations: Self-Reliance in China and North Korea." in *Culture and International Relations*, ed. Jongsuk Chay, 253-66. New York: Praeger, 1990.

Parris, Kristen. "Taxing the Private Sector in China: Implementing Policy Reform." Paper presented at the Western Political Science Association, Pasadena California, March 1993.

Potter, Sulamith H. and Jack M. Potter. *China's Peasants: The Anthropology of a Revolution*. Cambridge: Cambridge University Press, 1990.

Pressman, Jeffrey L. and Aaron Wildavsky. *Implementation*. Berkeley: University of California Press, 1984.

Pye, Lucian W. *Asian Power and Politics* (Cambridge: Belknap Press, 1985), p.20.

______. The Dynamics of Factions and Consensus in Chinese Politics: A Model and Some Propositions. Santa Monica, CA: Rand, 1980.

______. *The Mandarin and the Cadre: China's Political Cultures*. Ann Arbor: University of Michigan Press, 1988.

______. *The Spirit of Chinese Politics*. Cambridge: Harvard University Press, 1992.

Redclift, Michael. *Sustainable Development: Exploring Contradictions*. London: Routledge, 1987.

Ricoeur, Paul. *Hermeneutics and the Human Sciences: Essays on Language, Action and Interpretation*, ed., trans., J. Thompson. New York: Cambridge University Press, 1981.

______. *Interpretation Theory: Discourse and the Surplus of Meaning*. Ft. Worth: The Texas Christian University Press, 1976.

Rochester, J. Martin. "The Rise and Fall of International Organization as a Field of Study." *International Organization*. 40 (Fall 1986): 777.

Rosenau, James N. *Turbulence in World Politics: A Theory of Change and Continuity*. Princeton: Princeton University Press, 1990.

Rubinstein, Robert. "Cultural Analysis and International Security." *Alternatives* 13 (1988): 529-42.

Scarborough, Milton. *Myth and Modernity*. Albany: SUNY Press, 1994.

Schneider, Anne and Ingram, Helen. "Behavioral Assumptions of Policy Tools." *Journal of Politics* 52 (1990): 510-29.

Shambaugh, David. *Beautiful Imperialist: China Perceives America*. Princeton: Princeton University Press, 1991.

Shapiro, Michael J. *Language and Politics*. New York University Press, 1984.

______. *Language and Political Understanding: The Politics f Discursive Practices*. New Haven: Yale University Press. 1981.

______. *Reading the Post-Modern Polity*, University of Minnesota Press, 1992.

Shichor, Yitzak. "China and the Gulf Crisis: Escape from Predicaments." *Problems of Communism* 40 (Nov.-Dec. 1991): 78-90.

Shirk, Susan. "Playing to the Provinces: Deng Xiaoping's Political Strategy of Economic Reform." *Studies in Comparative Communism* 23 (Fall/Winter 1990).

Sicular, Terry. "China's Agricultural Policy During the Reform Period." In *China's Economic Dilemmas in the 1990's: The Problems of Reforms, Modernization, and Interdependence*, Joint Economic Committee, Congress of the United States, eds., 340-364. London: M.E. Sharpe, 1992.

Siu, Helen F. *Agents and Victims in South China: Accomplices in Rural Revolution*. New Haven: Yale University Press, 1989.

Smil, Vaclav. "Land Degradation in China: An Ancient Problem Getting Worse." in *Land Degradation and Society*, eds., Blaikei, Piers, and Harold Brookfield., 185-220. New York: Routledge, 1987.

Smircich, Linda. "Concepts of Culture and Organizational Analysis." *Policy Sciences* 19 (1986): 41-59.

Smith, Roger K. "Explaining the Non-Proliferation Regime: Anomalies for Contemporary International Relations Theory." *International Organization* 41 (Summer 1989): 349.

Snidal, Duncan. "Limits of Hegemonic Stability Theory. "*International Organization* 39 (Fall 1985): 579.

Spradley, James P. *The Ethnographic Interview*. New York: Harcourt Brace Jovanovich College Publishers, 1979.

Statistical Yearbook of China. English Edition. Oxford: Oxford University Press, 1985.

______. English Edition. Oxford: Oxford University Press, 1994.

Steinberg, David I. "The Confucian Backdrop: Setting the Stage for Economic Development." in *Multinational Managers and Host Government Interactions*. ed., Lee A. Tavis, 73-101. Notre Dame: University of Notre Dame Press, 1988.

Sun Dunli, Li Chaohai, Ma Xinming, and Wu Dafu. "Improving Eco-Environment and Enhancing Agricultural Sustaining Development." In *Integrated Resource Management For Sustainable Agriculture: Proceedings of the International Conference*, eds., Shi Yuanchun and Cheng Xu, 479-485. Beijing: Beijing Agricultural University Press, 1994.

Taylor, Charles. *Human Agency and Language: Philosophical Papers 1*. Cambridge: Cambridge University Press, 1985.

______. *Multiculturalism and The Politics of Recognition*, Princeton University Press, 1992.

Thomas, John W., and Merilee Grindle. "After the Decision: Implementing Policy Reforms in Developing Countries." *World Development* 18 (1990): 1163-1181.

______. "Policy Makers, Policy Choices, and Policy Outcomes: The Political Economy of Reform in Developing Countries." *Policy Sciences* 22 (1989): 213-248.

Thompson, John B. *Ideology and Modern Culture: Critical Social Theory in the Era of Mass Communication*. Stanford: Stanford University Press, 1990.

Thompson, Michael, Richard Ellis, and Aaron Wildavsky. *Cultural Theory*. Boulder: Westview Press, 1990.

Tso, T.C., ed. *Agricultural Reform and Development in China*. Beltsville, Md.: Ideals, Inc., 1990.

Tuchman, Barbara, W. *Stilwell and the American Experience in China: 1911-1945*. New York: Bantam Books, 1971.

Unger, Jonathan. "Bending the School Ladder: The Failure of Chinese Educational Reform in the 1960's." *Comparative Education Review* (June 1980): 221-237.

Van Ness, Peter. "China as a Third World State." In *China's Search for National Identity*, eds., Lowell Dittmer and Samuel Kim, 194-214. Ithaca: Cornell University Press, 1993.

Viotti, Paul R. and Kauppi, Mark V. *International Relations Theory*. New York: Macmillan Publishing Company, 1987.

Walker, R.B.J. "The Concept of Culture in the Theory of International Relations." in *Culture and International Relations*, ed. Jongsuk Chay, 4-17. New York: Praeger, 1990.

Waltz, Kenneth. "Anarchic Orders and Balances of Power." In *Neorealism and Its Critics*, ed., Robert Keohane, 98-130. New York: Columbia University Press, 1986.

______. "Political Structures." In *Neorealism and Its Critics*, ed., Robert Keohane, 70-97. New York: Columbia University Press, 1986.

______. "Reductionist and Systemic Theories." In *Neorealism and Its Critics*, ed., Robert Keohane, 47-69. New York: Columbia University Press, 1986.

Wang Zhengping, Li Jingling. "The Key of Sustainable Agriculture is to Develop Efficient Type Agriculture." In *Integrated Resource Management For Sustainable Agriculture: Proceedings of the International Conference*, eds., Shi Yuanchun and Cheng Xu, 574-579. Beijing: Beijing Agricultural University Press, 1994.

Wang Zhaoqian. "Historical Cultural and Social Backgrounds of Chinese Ecological Agriculture." Paper presented at the World Sustainable Agriculture Association Conference. Beijing, People's Republic of China, September 1994.

Wasserstrom, Jeffrey N., and Elizabeth J. Perry, eds. *Popular Protest and Political Culture in Modern China: Learning from 1989*. Boulder: Westview Press, 1992.

Watson, James L. "Rites of Beliefs." In *China's Search for National Identity*, eds., Lowell Dittmer and Samuel Kim, 80-103. Ithaca: Cornell University Press.

Wendt, Alexander. "The Agent-Structure Problem in International Relations Theory." *International Organization* 41 (Summer 1987): 335-70.

______. "Anarchy is What States Make of it: The Social Construction of Power Politics." *International Organization* 46 (1992): 391-425.

______. "Collective Identity Formation and the International State." *American Political Science Review* 88, no. 2 (June 1994): 384-96.

______. "Levels of Analysis vs. Agents and Structures." *Review of International Studies* 18 (1992): 181-5.

______. "Hierarchy under Anarchy: Informal Empire and the East German State." *International Organization* 49, no. 4 (Autumn 1995): 689-721.

White, Lynn, and Li Cheng. "China Coast Identities." In *China's Search for National Identity*, eds., Lowell Dittmer and Samuel Kim, 154-93. Ithaca: Cornell University Press, 1993.

Whyte, William Foote. *Learning From the Field: A Guide From Experience*. London: Sage Publications, 1984.

Wilhelm, Richard, translator. *The I Ching or Book of Changes*. Princeton: Princeton University Press, 1985.

Williams, C.A.S. *Outlines of Chinese Symbolism and Art Motives*. New York: Dover Publications, 1976.

Wilson, Richard. "Change and Continuity in Chinese Cultural Identity: The Filial Ideal and the Transformation of an Ethic." In *China's Quest for National Identity*, eds., Lowell Dittmer and Samuel Kim, 104-124. Ithaca: Cornell University Press, 1993.

Wolcott, Harry F. *Transforming Qualitative Data: Description, Analysis, and Interpretation*. London: Sage Publications, 1994.

Xu, Cheng, Han Chunru, and Donald S. Taylor. "Sustainable Agriculture Development in China." *World Development* 20 (August 1992): 1127-1143.

Yang, Chung-fang. "Conformity and Defiance on Tiananmen Square: A Social Psychological Perspective." In *Culture and Politics in China: the Anatomy of Tiananmen Square*. eds. Peter Li, Steven Mark, and Marjorie H. Li, 197-224. New Brunswick, N.J.: Transaction Publishers, 1991.

Yang Minchuan. "Reshaping Peasant Culture and Community: Rural Industrialization in a Chinese Village." *Modern China* 20 (April 1994): 157-179.

Yang Wenjing. "Study on Eco-economic Management Policy under Market Economy System." *Sheng Tai Jing Ji* [Ecological Economics] 2 (March 1994): 10-16.

Yanow, Dvora. "The Communication of Policy Meanings: Implementation as Interpretation and Text." *Policy Sciences* (Spring 1994): 41-59.

______. "Reflecting on Methods: Garbage-can Cases and Footholds of the Mind." In *How Does a Policy Mean*. Unpublished manuscript, 1994.

______. "Silences in Public Policy Discourse: Organizational and Public Policy Myths." *Journal of Public Administration Research and Theory* 2 (October 1992): 399-423.

______. "Supermarkets and Culture Clash: The Epistemological Role of Metaphors in Administrative Practice." *The American Review of Public Administration* 22 (June 1992): 89-110.

Yin, Robert K. *Case Study Research: Design and Methods*. London: Sage Publications, 1989.

Young, Oran R. "International Regimes: Problems of Concept Formation." *World Politics* (1980): 331.

______. "The Politics of International Regime Formation: Managing Natural Resources and the Environment." *International Organization* 43 (Summer 1989): 377.

Zacker, Mark W. "Trade Gaps, Analytic Gaps: Regime Analysis and International Commodity Trade Regulation." *International Organization* 41 (Spring 1987): 253.

Zhang Renwu, Ji Wenying, and Zhang Tong. "Analysis on the Characteristics and Components of China's Sustainable Agriculture Technological System." Paper Presented at the World Sustainable Agriculture Association Conference. Beijing, 1994.

Zhang Shuqiang. "Marxism, Confucianism, and Cultural Nationalism." In *The Chinese and Their Future*, eds., Zhiling Lin and Thomas W. Robinson, 82-109. Washington D.C.: AEI Press, 1994.

Zhang Taolin, Zhao Qiguo, Zhang Bin. "On Sustainable Agricultural Development in South China." In *Integrated Resource Management For Sustainable Agriculture: Proceedings of the International Conference*, eds., Shi Yuanchun and Cheng Xu, 386-392. Beijing: Beijing Agricultural University Press, 1994.

Zhang Xigu. "The Key to the Breakthrough for Chinese Sustainable Agriculture: Organic Combination of Chinese Traditional and Modern Agricultural Techniques." Paper presented to the World Sustainable Agriculture Association. Beijing, People's Republic of China, September 1994.

Zhong, Funing, Jiang Shirong, and Wang Rong. "Preliminary Analysis of Seabeach Land Development in Northern Jiangsu Province." Unpublished manuscript, 1994.

Zhou Shengkun. "A Farmer First Paradigm in Order to Complement Transfer of Technology Tendency Towards Sustainable Agriculture in China in the Nineties." In *Integrated Resource Management For Sustainable Agriculture: Proceedings of the International Conference*, eds., Shi Yuanchun and Cheng Xu, 653-60. Beijing: Beijing Agricultural University Press, 1994.

Zhu Ling. "The Transformation of the Operating Mechanisms in Chinese Agriculture." *The Journal of Development Studies* 26 (January 1990): 231.

Zweig, David. "Peasants and Politics." *World Policy Journal* 64 (Fall 1989): 620-645.